A Specialist Periodical Report

Surface and Defect Properties of Solids
Volume 3

A Review of the Recent Literature Published up to April 1973

Senior Reporters
M. W. Roberts, *School of Chemistry, Bradford University*
J. M. Thomas, *Edward Davies Chemical Laboratories, University College of Wales, Aberystwyth*

Reporters
P. S. Allan, *Liverpool University*
J. S. Anderson, *Oxford University*
M. J. Bevis, *Liverpool University*
A. M. Bradshaw, *Technischen Universität, München*
W. D. Erickson, *NASA Langley Research Centre, Hampton, Virginia*
W. F. Harris, *Witwatersrand University*
J. W. Linnett, *Cambridge University*
R. A. Suthers, *Cambridge University*
R. J. D. Tilley, *Bradford University*
G. Webb, *Glasgow University*

© Copyright 1974

The Chemical Society
Burlington House, London, W1V 0BN

ISBN: 0 85186 270 5
Library of Congress Catalog Card No. 72-78528

*Printed in Great Britain by Billing & Sons Limited
Guildford and London*

Preface

With the publication of this, the third volume of 'Surface and Defect Properties of Solids', it is pertinent to reflect on progress and development in this area of chemistry over the past five years. There is little doubt that we are currently in a period of rapid and profound development.

In Volume 1 Anderson presented a definitive review of Shear Structures and Non-stoicheiometry and, in so doing, dealt in particular with the natural emergence of the notion of extended defects from the confused background of theoretical treatments which were formulated within the restricted and restrictive concept of point defects. Since that time, chiefly as a result of the application of high-resolution (lattice-fringe) electron microscopy to systems which possess simplifying structural features and, equally importantly, considerable stability under high-flux electron irradiation, there has been feverish activity in the study of shear- and block-structures. Unexpected subtleties, such as 'swinging' and 'rotary' crystallographic shear and seemingly infinitely adaptive block structures, have come to light, and new unifying principles have been uncovered. As yet the full theoretical justification for interpreting lattice-fringes [recorded under well-defined conditions of defects-of-focus, crystal thickness and the precise number of $(n-)$ beams] has not been forthcoming; and much remains to be learned about the characteristics of the bonding in these structures. The position reached in this burgeoning area of solid-state chemistry is covered in Chapter 1 by Anderson and Tilley, who have pursued the policy prevalent in our previous volumes, namely to be selective and critical rather than exhaustive and comprehensive. The topic of grossly defective solids is expanding rapidly and will doubtless require further review in a few years time.

Disclinations, the subject of the report by Harris, who has contributed much to the elucidation of the properties of such defects, have been demonstrated, as is made clear in Chapter 2, to be of considerable significance in solids or quasi-crystalline materials where the state of organization and order occurs at a higher level than the simple atomic or molecular scale. It is clear from this review that disclinations frequently govern the properties of aggregated materials, especially those relevant in biological and biophysical phenomena.

One of the main barriers that thwarts progress in unravelling the ultramicrostructural characteristics of organic solids is general lack of stability, in electron-beams, of molecular crystals, the accurate unit-cell dimensions of which have already been previously established by conventional X-ray and neutron diffraction. Some typical questions that need to be answered include: whether there are extensive two-dimensional defects in such solids; or whether dislocations are dissociated into ribbons, or whether they may be readily rendered mobile and generate debris of various kinds during motion. Answers are, at present, not readily available. But some solids, including those which are not particularly electron beam-resistant (*e.g.* polyethylene) have been amenable to specialized electron-optical techniques and have yielded unexpectedly interesting phenomena. The phenomenon of stress-induced phase transformation ranks amongst these and it highlights the question of phase-transformation and polymorphism generally. Bevis and Allan give a detailed description of the phenomenology and interpretation of such transformations in Chapter 3, which illustrates well the advantages that the approaches of the physicists and metallurgists can offer the solid-state chemist.

The attention given to the surface in Volumes 1 and 2 has emphasized the new approaches being developed around photoemission and spectroscopy, and in this volume the chapter by Bradshaw deals with the particular subject of Appearance Potential Spectroscopy. Joyner and Somorjai reviewed the recent trends in LEED application in Volume 2, but there is as yet no clear indication that the theoretical problems associated with the interpretation of LEED data have been overcome, although claims of varying degrees of success have recently been made. Central to the problems of surface crystallography is the question of the electronic structure of the surface molecule, and in Chapter 4 Linnett and his colleagues discuss the application of the FSGO approach to a number of relatively simple structures. These include Li_nH_n (clusters), LiH (crystal), and the adsorption of He on the (100) face of LiH (crystal).

In the field of catalysis we have ranged from one of the simplest of catalytic reactions, $H_2 + D_2$ (Gasser, Volume 1), to more complex ones involving saturated hydrocarbons (Kempling and Whan, Volume 2). In this volume Webb considers some aspects of unsaturated hydrocarbons, emphasizing the complexities involved and at the same time giving a clear lead as to where progress is being made.

We currently have a plethora of experimental techniques available and their successful application will revolve around whether the process of investigation distorts or destroys the surface. It is one thing to make an observation, it is another to relate it to the situation at the surface prior to the investigation. There are now well-recognized restrictions to Auger Electron Spectroscopy and question marks remain with techniques such as Secondary Ion Mass Spectrometry (SIMS) which are as yet in their early stages of development. Considerations such as these will be discussed in Volume 4.

M. W. R.
J. M. T.

Contents

Chapter 1 Crystallographic Shear and Non-stoicheiometry   1
By J. S. Anderson and R. J. D. Tilley

  1 Introduction   1

  2 The Direct Observation of Structure in Crystals: Lattice Imaging   2
    Theoretical Interpretation of Image Contrast   4
    Experimental Observations of Lattice Images   11
    Electron Microscopy and Lattice Imaging in the Study of Reactions   21

  3 Rotating CS Planes   25
    CS Structures in the Titanium and Titanium–Chromium Oxides   25
    CS Structures in the Tungsten Oxides   26
    The Topology of CS Structures   26
      ReO_3-related Structures   27
      Rutile-related Structures   29
      α-PbO_2-related Structures   32
      Rotary Crystallographic Shear   34

  4 Block Structures   36
    The Topology of Block Structures   36
    Non-stoicheiometry and Defects in Block Structures   40
    Non-stoicheiometric Block Structures   42

  5 Infinitely Adaptive Structures   49
    CS Phases   49
    Adaptive Superlattice Ordering   50

Chapter 2 The Geometry of Disclinations in Crystals   57
By W. F. Harris

  1 Introduction   57
    Historical Remarks   61

2 Weingarten–Volterra Dislocations in Isotropic Bodies 63
Dispirations and Disclinations 64
Components of a General Dispiration 66
Limitations on Rotations of Disclinations and
  Dispirations 68

3 Disclinations and Dispirations in Structured Materials 70
Disclinations in Crystals 70
Dispirations in Crystals 72
Defects in Surface Crystals 73
  Intrinsic Disclinations 75
  Extrinsic Disclinations and Dispirations 79
  Closed Surface Crystals 80
Disclinations in Other Materials 86

4 Movement of Disclinations in Solids 89

Chapter 3 Stress-induced Martensitic Transformations and Twinning in Organic Molecular Crystals 93
By M. J. Bevis and P. S. Allan

1 Introduction 93

2 The Crystallography of Martensitic Transformations 99
Martensitic Transformations in Crystalline Polymers 102
The Crystallography of 'Two-dimensional' Martensitic
  Phase Transformations 103
The Crystallography of 'Two-dimensional' Deformation
  Twinning 106
Comparison of Theory and Experiment 106

3 Application of the Martensite and Twinning Crystallography Theories 107

4 Experimental Procedures used in Electron Diffraction and Microscopy Studies of Deformed Polymer Crystals 112

5 Experimental Results and Discussion 113

Contents

Chapter 4 A Simple Wavefunction for Solid and Surface Calculations   132
By R. A. Suthers, J. W. Linnett, and W. D. Erickson

   1 Introduction   132

   2 Li_nH_n Cluster Calculations   137

   3 The LiH Infinite Crystal   140

   4 The [100] Surface of Crystalline LiH   143

   5 Adsorption of Helium on to the [100] and [110] Surfaces of Crystalline LiH   146

   6 Conclusion   150

   7 Bibliography of FSGO Molecule Calculations   152

Chapter 5 Appearance Potential Spectroscopy and Related Techniques   153
By A. M. Bradshaw

   1 Introduction   153

   2 Underlying Principles of SXAPS   157
     Background Effects   157
     The Intensity of APS Features   158

   3 Some Experimental Considerations   162
     Apparatus   162
     Determination of the Threshold Energy   165

   4 Applications   166
     Surface Chemical Analysis   166
     Binding Energies in the Third Transition Metal Series and Band Structure   168
     Carbon and Plasmon Coupling Phenomena   172
     Simple Metals   176
     Alloys   177
     Chemical Shifts   180

   5 Summary   181

Chapter 6 Some Aspects of the Nature and Reactivity of
  Adsorbed States of Unsaturated Hydrocarbons
  on Metal Catalysts   184
  By G. Webb

  1 Introduction   184

  2 Nature of the Adsorbed States of Unsaturated Hydrocarbons   184

  3 Infrared Spectra of Adsorbed Hydrocarbons   188

  4 Surface Migration and the Influence of Catalyst Supports   192

  5 The Influence of Catalyst Structure on Reactivity   196

Erratum   198

Author Index   199

1
Crystallographic Shear and Non-stoicheiometry

BY J. S. ANDERSON AND R. J. D. TILLEY

1 Introduction

When the previous Report in this series was written, detailed experimental evidence about the microstructure of crystallographic shear structures (CS phases) was just becoming available. Two themes were therefore emphasized. The first was the relation between the concept of crystallographic shear and existing views of defects and non-stoicheiometry in inorganic compounds. Most CS phases do not appear to contain point defects in significant concentrations—*i.e.* in sufficient number to contribute materially to the apparent composition ranges of CS compounds. In a heuristic sense, at least, the collapse of the parent crystal structure which produces a CS plane eliminates point defects; it is not implied that the presence of point defects in high concentration is a necessary precursor stage in the formation of a CS plane. The mechanism of this transformation process is still not clear, and the role of point defects in certain structural types, notably the 'block' structure oxides, has continued to excite interest.

A second topic, which has also been actively pursued in the review period, was the importance of coherent intergrowth between structures that are topologically compatible, but of different composition, exhibited particularly by the CS phases. It was shown that in many macroscopically homogeneous CS phases there may be a considerable measure of internal disorder, associated with irregularities in the spacing between parallel CS planes. Even though this state may not represent a true equilibrium structure, it may be experimentally inescapable and it provides a basis for apparent non-stoicheiometric properties at the macroscopic level. The usual methods of characterization and structure analysis may fail to reveal and analyse such microscopic heterogeneity; to do so needs methods for determining the local microstructure, at the unit-cell level, as distinct from the averaged structure derived from diffraction methods and from postulated models for defect structures. Such methods were beginning to emerge from the application of electron microscopy.

In the intervening two years, the power of lattice-imaging methods in electron microscopy has developed markedly. Considerable attention has been devoted to coherent intergrowth (Wadsley defects) and other forms of faulting, as observed at or below the unit-cell level, and the topological constraints and relationships that determine the possibilities of intergrowth or structural relaxation in CS phases have been analysed. This advance in our knowledge of structure has not been matched by advances in knowledge of transport and reaction processes, and the physical properties associated with CS structures have received little study. In this review, we consider particularly the increasingly important role of electron microscopy, and the way in which structure adjusts itself to composition, both in materials that simulate non-stoicheiometric behaviour and those that are genuinely variable in composition. Crystallographic shear—the term has a very specific meaning, which should not be loosely used—is not the only transformation whereby inorganic structures can accommodate changes in the atomic ratio of metal : non-metal so as to maintain some high degree of order. Recent work has drawn attention to the formation of ordered structures, with large repeating units, where randomized solid solutions might be expected. Without venturing an answer to the difficult questions of how a solid compound should now be defined, or how complex ordering is established, we summarize also some recent developments in this wider field.

2 The Direct Observation of Structure in Crystals: Lattice Imaging

The newer experimental findings about CS phases have come largely from transmission electron microscopy and electron diffraction, rather than from X-ray diffraction, which has limitations imposed by the large unit cells, by the small differences in structure between one member and another in a homologous series, and, above all, by disorder in the crystals. The first results accruing from electron microscopy[1] indicated that complete order in crystals of CS phases was rarely attained; it may, indeed, be impossible even by the most careful preparative methods to obtain perfectly ordered crystals of many CS compounds for structure determination by single-crystal methods.

Transmission electron microscopy can present the essentials of the structure of CS phases and other suitable classes of compound very directly, although detailed metrical information—*e.g.* interatomic distances—cannot be extracted. For this, there is no alternative to precise structure determination by X-ray or, increasingly and for some purposes advantageously, neutron-diffraction methods. Lattice images can now be obtained, however, which show the projected structure at the level of the individual co-ordination polyhedron in oxide structures. Development of the technique has been largely pragmatic, based on the experimental finding that, provided that the crystals under examination were extremely thin, the contrast in micrographs made at 'optimum under-focus' approximated closely to a projection of the

---

[1] J. S. Anderson, 'Surface and Defect Properties of Solids', ed. M. W. Roberts and J. M. Thomas (Specialist Periodical Reports), The Chemical Society, London, 1972, Vol. 1, p. 1.

potential distribution in the crystal.[2] Thus, for CS phases, in which the density of heavy cations, of high charge, is considerably higher in the CS planes than in the relatively open matrix of parent structure, the CS planes appear as fringes which collapse into dark lines of contrast when the crystal is so oriented as to bring the CS planes parallel to the incident beam. At the highest lattice resolution, individual corner-sharing (MO_6) octahedral groups appear darker than the empty voids between them.

Interpretation of these lattice images is, however, not as straightforward as might appear at first sight. Ideally, an electron micrograph should be compared, point for point, with the intensity of the transmitted electron beam, as calculated from dynamical scattering theory. In practice, the theory of an electron-microscope image formed by the operation of many diffracted beams has been developed only in parallel with the experimental applications. Most results on CS and other structures have relied upon a less rigorous comparison with electron optic theory; much of the interpretation has, indeed, come from chemical intuition regarding the structural geometry that might be expected in the system concerned.

The correspondence between lattice images and the structures postulated from a combination of X-ray structural information with chemical intuition has been remarkably good, and the validity of the interpretations has not been in serious doubt. This success has stemmed from careful selection of crystalline systems, with a known basic structure, that were appropriate for attack by microscopy. The electron-microscope image could then be so focused as to give optimum contrast and resolution which brought out structural features that harmonized with expectations based on the prior knowledge of the system. Aperiodic features of the image, faulting, disorder, *etc.* could then be interpreted in a manner consistent with the interpretation in perfect regions of crystal. The rather open block structures were particularly suitable for these tactics.

If the lattice-imaging technique is to be extended to the study of crystals which have a less open and clearly projected structure, a rigorous theoretical basis for interpretation becomes indispensible. Lattice images at a sufficiently high degree of resolution to give some chance of displaying the positions of individual, highly scattering atoms (*ca.* 0.35 nm with current instruments, compared with the cation–cation distance 0.39 nm between apex-sharing octahedral groups) have hitherto been available only for the block structures. These have, accordingly, been the objects for calculations of image contrast as the electron optic theory has developed during the review period. The calculations have a much wider validity, however, for they define the conditions under which lattice images may be directly correlated with structure and thus now permit the technique to be employed for a wider range of chemically interesting systems. We therefore briefly discuss the relation between the recorded lattice image and the structure, seen in projection, of

---

[2] J. G. Allpress and J. V. Sanders, *J. Appl. Cryst.*, 1973, **6**, 165.

the observed crystal, before considering areas in which electron microscopy has shed new light on defect structures and chemical problems.

Theoretical Interpretation of Image Contrast.—In general terms, contrast in a micrograph reflects variations in the numbers of electrons falling upon the recording photographic emulsion; in an ideal optical system, the electrons arriving at a single point on the emulsion originate from a single point on the exit face of the object. The flux at each point on the emulsion depends upon the diffracting conditions within the crystal, upon instrumental factors, and upon operating factors. It is therefore possible to break down the theoretical problem of image formation from a large number of diffracted beams into three distinct parts: (*a*) calculation of the electron wavefunction at the exit face of the crystal; (*b*) modification of the wavefunction through the optical system, to take account of lens aberrations, moveable apertures, *etc.*; and (*c*) superimposed effects of operating adjustments and errors. Apart from the critical control of the orientation of the specimen, the major factor in (*c*) is the extent of under-focus; this is selected to compensate for the phase incoherence produced by spherical aberration and to give the best match between the observed image and that calculated theoretically.

Experimental experience has shown that useful lattice images are produced only from very thin crystals; even for crystals a few nanometres thick, realistic values for the intensity and phases of the various diffracted beams can be calculated only by using dynamical diffraction theory.[3] When lattice fringe images are formed by the combination of more than one or two diffracted beams (for which the electron optic theory is well established), the dynamical formulation of Cowley and Moodie[4] provides the most useful starting point for calculations.

The essence of this treatment is to consider the crystal to be divided into a number of thin slices perpendicular to the incident beam direction. The phase and amplitude of the incident wave is then modified by each slice. This is further simplified by considering each slice as a phase grating formed by projecting the potential distribution within each slice on to an internal plane. Fresnel diffraction then takes place between each grating. The treatment becomes less rigorous with increasing slice thickness, but provided this is kept small, no appreciable errors are found.[5] Thus, each time the electron wave passes across a slice, it is multiplied by a transmission function which is dependent upon the potential distribution in the plane.

If the potential of the nth slice, of thickness Δz, is $\phi_n(x,y,z)$ then the projected potential, $\phi_n(x,y)$ is given by

$$\phi_n(x,y) = \int_z^{z+\Delta z} \phi_n(x,y,z)\,\mathrm{d}z \qquad (1)$$

[3] J. M. Cowley, 'Progress in Materials Science', ed. B. Chalmers and W. Hume-Rothery, Pergamon, Oxford, 1967, Vol. 13, p. 267.
[4] J. M. Cowley and A. F. Moodie, *Acta Cryst.*, 1957, **10**, 609.
[5] G. R. Grinton and J. M. Cowley, *Optik*, 1971, **34**, 221.

The transmission function for the slice, the phase grating, is then defined as

$$q(x,y) = \exp(i\sigma\phi_n) \qquad (2)$$

and gives the phase change imposed upon the wave as it passes the grating. The term in σ is a function of the relativistic electron wavelength λ of velocity v and the accelerating potential W, given by

$$\sigma = \frac{\pi}{\lambda W} \frac{2}{1 + (1 + \beta^2)^{1/2}} \qquad (3)$$

where β is the relativistic correction term, v/c. The phase change between the planes must also be considered. This can be represented as $\exp[ik(x^2 + y^2)/2\Delta z]$ for the fast electrons that are involved. Adding these phase changes, by use of the principle of superposition, allows the wavefunction of the $(n+1)$th slice to be written in terms of the nth slice: thus

$$\psi_{n+1} = \left\{\psi_n * \exp\left[\frac{ik(x^2 + y^2)}{2\Delta z}\right]\right\} \exp(i\sigma\phi_{n+1}) \qquad (4)$$

where $*$ represents convolution. The final wavefunction emerging from the crystal is obtained from equation (4) and an analytical solution can be derived.[4] The form this takes, however, is rather unsuitable for numerical computation, and an iterative scheme based on the equation for ψ_{n+1} has been used.[5-9]

In order to do this, it is convenient to handle the Fourier transform of equation (4), which is

$$U_{n+1}(h,k) = [U_n(h,k) \cdot P(h,k)] * Q_{n+1}(h,k) \qquad (5)$$

where $U_n(h,k)$ is the wave amplitude and phase from the nth slice, P is the propagation function

$$P(h,k) = \exp[2\pi i\zeta(h,k)\Delta z] \qquad (6)$$

where $\zeta(h,k)$ is the excitation error, $(u^2 + v^2)\lambda/2$, for the (h,k) reflection in the reciprocal space co-ordinates (u,v), and $Q_{n+1}(h,k)$ is the Fourier transform of the phase grating function, $q_{n+1}(x,y)$ given by

$$Q(h,k) = \int_0^a\int_0^b q(x,y)\exp\left[2\pi i\left(h\frac{x}{a} + k\frac{y}{b}\right)\right]\mathrm{d}x\,\mathrm{d}y \qquad (7)$$

The equation in U, equation (5), is evaluated for a sufficient number of slices to give the correct crystal thickness, and a sufficient number of beams so that the sum of the intensities is close to the incident-wave intensity. In the calculations referred to,[6-9] 435 beams were used in two-dimensional

---

[6] J. G. Allpress, E. A. Hewat, A. F. Moodie, and J. V. Sanders, *Acta Cryst.*, 1972, **A28**, 528.
[7] D. F. Lynch and M. A. O'Keefe, *Acta Cryst.*, 1972, **A28**, 536.
[8] G. R. Anstis, D. F. Lynch, A. F. Moodie, and M. A. O'Keefe, *Acta Cryst.*, 1973, **A29**, 138.
[9] M. A. O'Keefe, *Acta Cryst.*, 1973, **A29**, 389.

calculations. Normalization was 0.72 for a crystal thickness of 60 nm. This would be improved by including more beams, but at the expense of much greater computation time. The final result is to generate the diffraction pattern, which, because of the variable nature of Δz, is obtained as a function of thickness.

The contrast in the image is readily derived from the function U_n by another Fourier transform. If ψ_H is the Fourier transform of U_H so that

$$\psi_H(x,y) = F^{-1} U_H(h,k) \tag{8}$$

the image intensity is given by

$$I(x,y) = \psi(x,y) \cdot \psi * (x,y) \tag{9}$$

where F^{-1} is the inverse Fourier transform of $U_H(h,k)$, the amplitude and phase of the (h,k)th diffracted beam emerging from the crystal of thickness H. The representation of this image can be expressed in a variety of ways. The most immediately informative, however, is to plot out the expected contrast, using the half-tone print-out process developed by Head,[10] to give a computed micrograph.

These results relate to a perfect, aberration-free electron microscope, and they now have to be modified to take into account the instrumental and operational defects mentioned earlier. These are principally the inclusion of terms to account for the objective aperture, the spherical aberration of the objective lens, and the defect of focus employed. The objective aperture is represented by merely excluding all beams intercepted by the objective aperture from the final Fourier transformation. The spherical aberration acts so as to retard the phase of a beam passing through the lens at an angle α to the optical axis by an amount $\pi C_s \alpha^4 / 2\lambda$ with respect to the axial beam at the Gaussian image plane. C_s is the spherical aberration coefficient of the objective lens, which lies between 3 and 5 mm for most modern instruments. The defect of focus ε is taken into account by including a propagating function which is convoluted with the wavefunction at the exit surface of the crystal to allow the wavefunction at a distance ε from the exit surface to be obtained. Thus equation (8) becomes modified to

$$\psi_{H,\varepsilon}(x,y) = F^{-1}\{U_H(h,k) \cdot P_\varepsilon(h,k)\} \tag{10}$$

where P is a propagating function which accounts for defect of focus and spherical aberration, and is given by

$$P_\varepsilon(h,k) = \exp\{2\pi i \zeta(h,k)[\varepsilon - \lambda C_s \zeta(h,k)]\} \tag{11}$$

for a defect of focus ε and spherical aberration coefficient C_s. $\zeta(h,k)$ is the

---

[10] A. K. Head, *Austral. J. Phys.*, 1967, **20**, 557.

excitation error for the (h,k)th beam. Hence we can write the image wave-function for an apertured system

$$\psi_{H,\varepsilon,n,C_s}(x,y) = \sum_h \sum_k U_H(h,k) \cdot \exp\left[2\pi i\left\{\zeta(h,k)[\varepsilon - \lambda C_s \zeta(h,k)] - \left(\frac{hx}{a} + \frac{ky}{b}\right)\right\}\right] \quad (12)$$

where the summations are carried out over the n beams supposed to pass through the aperture.

Other electron-microscope defects—chromatic aberration, the divergence of the incident beam, and astigmatism—can all be taken into account. However, these are found to be rather less important, and astigmatism in particular can be ignored for a carefully corrected modern objective lens. Equation (12) can then be used to compute an aberrated apertured image which in turn can be compared with the ideal one and with experimental results.

Calculations were first carried out for the case where only one row of diffracted beams, typically (h00 or 00l), is excited.[5] These give one-dimensional information, and a set of parallel fringes results, which can be compared with experimental lattice images obtained under conditions which are matched by the data used in the computations. Some of the results of these calculations are compared with the experimental results in Figure 1.

There are no theoretical problems associated with computing two-dimensional images using the same procedure.[8] Once again, half-tone images are the easiest way of comparing the results of computations with experimental micrographs. Some examples are shown in Figure 2.

From these results it is possible to see that under certain circumstances the lattice images do represent accurately the structure of the crystal or, to be more precise, the projected charge density of the crystal. However, the calculations show that such an interpretation is not generally valid. In fact one must place severe limitations on the experimental technique used in order to avoid misinterpretation of lattice images. The most important limitation is that the crystal must be very thin, of the order of 10 nm. Of course, if the crystal thickness is accurately known, computations of contrast can still be made, but the image contrast does not reflect the crystal structure in simple terms and cannot be interpreted naively. Another very important result is that the most important microscope aberration appears to be spherical aberration, which causes severe perturbations of image contrast and necessitates rather careful selection of the objective aperture used. This is allowed for by computing images which are produced by beams having considerable phase retardation due to spherical aberration. This emphasizes, however, that lattice images can be interpreted intuitively only when care has been taken in obtaining the micrographs. The final criterion which must be carefully controlled is the defect of focus. For the lattice image to represent

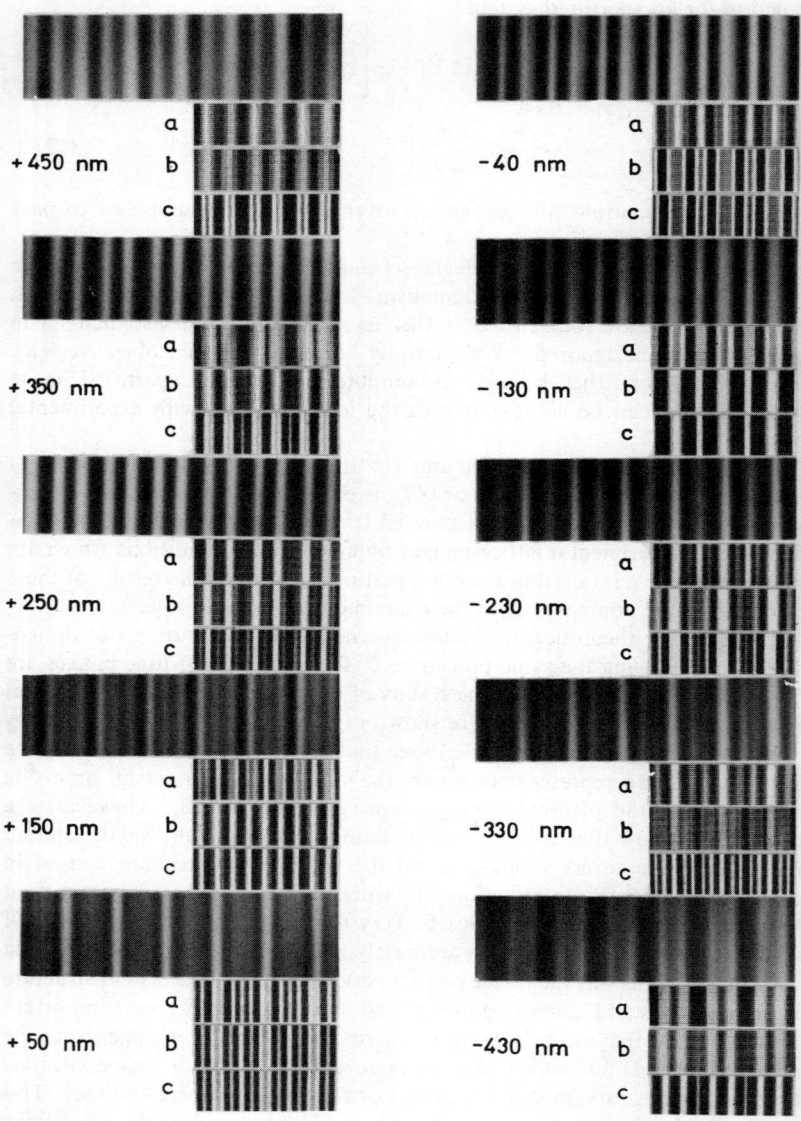

Figure 1 Comparison of an observed through-focus series (larger images) and computed images (smaller), showing the effect of various instrumental factors. The numbers refer to the defects of focus for which the images were computed: (a) corrected for divergence and spherical aberration; (b) corrected for divergence only; (c) 'ideal' image, with no corrections.
(Reproduced by permission from *Acta Cryst.*, 1972, **A28**, 528)

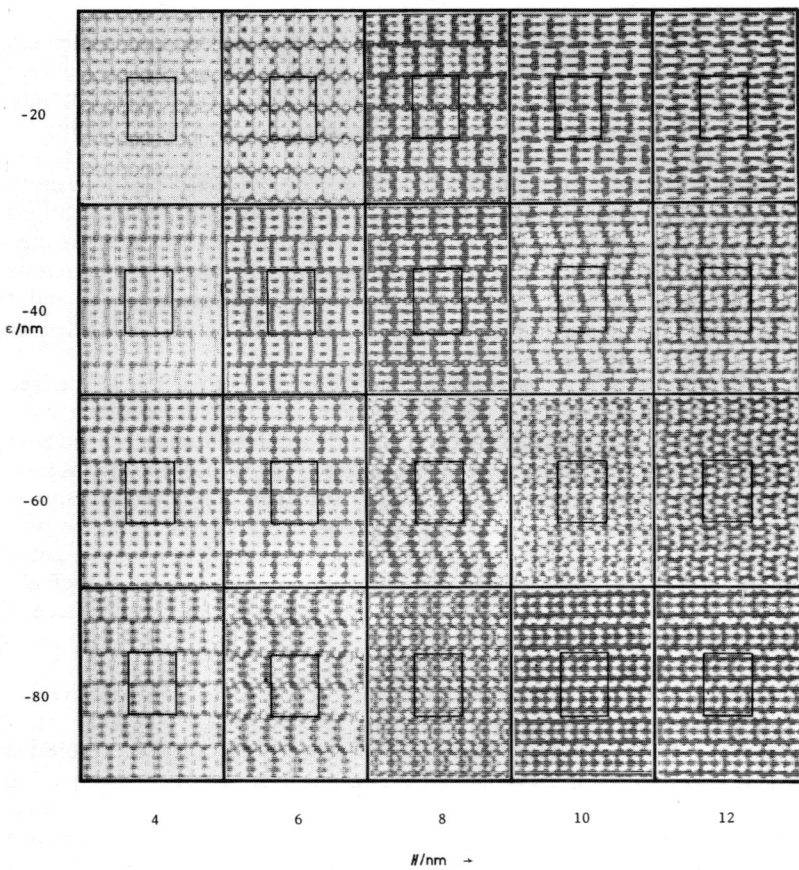

Figure 2 $Ti_2Nb_{10}O_{29}$ *n-beam lattice images showing the variation of contrast with crystal thickness, H/nm, and defect of focus ε/nm. The objective-lens spherical aberration coefficient, $C_s = 1.5$ mm.*
(Reproduced by permission from *Acta Cryst.*, 1973, **A29**, 389)

the crystal structure accurately, the microscope must typically be underfocused by about 80 nm. A degree of overfocus inverts the image contrast, but between these two positions, and outside them, the contrast varies in a rather complex way. Experimentally this is fairly easy to control, as the disappearance of Fresnel fringes at the edge of the crystal can be used to judge focus, and the correct defect of focus set by adjusting the objective lens by the appropriate amount. This is not necessary if the approximate crystal structure is known, for in that case the 'best' contrast can be judged by eye.

Such a procedure has been successfully used in the block structures, for example.

These results clearly show that it is highly desirable to compute images wherever possible. One problem of doing this is that the multislice computations are rather time-consuming. Some effort has thus been put into approximate computations which may be used to check the validity of structural models.[7]

The most useful of these approximations is the Projected Charge-density approximation, not only because of the relative simplicity of the calculations but also because of the attractive feature that image contrast can be related to charge density in the crystal on an intuitive basis. The experimental results and calculations have shown this to be possible in ideal circumstances, and it is therefore informative to compare calculations based on the charge-density model with the more accurate ones already described.

The approximation is based upon the theory of Fourier images derived by Cowley and Moodie.[11] This theory indicates that the periodic components of a two-dimensional phase object will produce a series of images at specific positions on either side of the object. In the electron microscope, these images have a magnification of approximately unity, and are spaced far enough apart for the normal focused image of the object to be the only one readily observed. A thin crystal can be considered as a simple phase object for the purposes of this approximation. The importance of the theory is that the intensity distribution in the image planes close to those of the Fourier images is given by[12]

$$I(x,y) = 1 - \frac{\varepsilon\lambda\sigma}{2\pi} \nabla^2 \phi(x,y) \qquad (13)$$

where I is the image contrast at defect of focus ε, $\phi(x,y)$ is the projected potential of the crystal, and λ and σ have been defined earlier. This equation holds for an unrestricted aperture and an idealized electron microscope. The potential distribution $\phi(x,y)$ is, however, closely related to the charge distribution in the crystal by Poisson's equation

$$\nabla^2 \phi_p(x,y) \propto \rho_p(x,y) \qquad (14)$$

Thus the image contrast for small values of ε can be written as

$$I(x,y) = 1 - k\varepsilon\rho_p(x,y) \qquad (15)$$

where k is a constant. Hence the image contrast is a direct representation of the projected charge density in the crystal.

This equation is applicable to both periodic and non-periodic defects, and can be derived directly from equation (8)[6] by taking the high-voltage limit of the scattering. In this case, the wavelength tends to zero and the Ewald sphere becomes planar. The images should reverse their contrast on passing

---

[11] J. M. Cowley and A. F. Moodie, *Proc. Phys. Soc.*, 1958, **71**, 533.
[12] J. M. Cowley and A. F. Moodie, *Proc. Phys. Soc.*, 1960, **76**, 378.

through the Gaussian focus. However, the spherical aberration of the objective lens perturbs the contrast and it is found that underfocused images correspond rather better with the charge-density distribution. This is, of course, in agreement with experimental observations. Figure 3 shows a

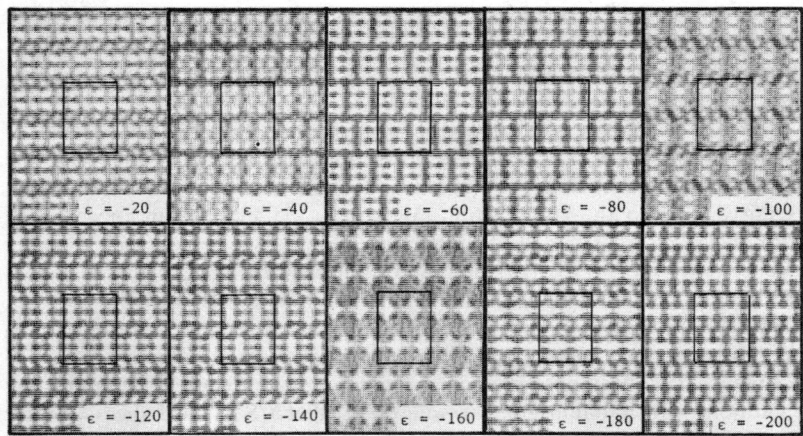

Figure 3 $Ti_2Nb_{10}O_{29}$ *n-beam lattice images for a crystal* 5 nm *thick at various values of defocus* ε/nm. *The objective-lens spherical aberration coefficient,* $C_s =$ 1.8 mm
(Reproduced by permission from *Acta Cryst.*, 1973, **A29**, 389)

series of lattice images computed from the charge-density approximation for various values of defocus. It is possible to conclude, from such results, that for very thin crystals, provided that the image contrast reverses on passing through focus, the charge-density interpretation is valid. In this case, then, image features can be naively interpreted in terms of structural features that produce a variation in charge density. The thickness at which this approximation fails will depend upon the structure under consideration and the electron-microscope operating voltage; higher voltages are desirable in that the experimental conditions then resemble the conditions employed in the approximation more closely.

Experimental Observations of Lattice Images.—The theoretical calculations indicate that for very thin crystals the image contrast should be a representation of the charge density in the flake. Thus at medium resolution (*ca.* 1 nm) CS planes should appear darker than the background whereas at higher resolutions they should appear broken into segments representing the groups of edge-shared octahedra in the CS plane. This will be so regardless of whether the material is ordered or disordered, and the main advantage over *X*-ray crystallography lies in the consequence that the local order, instead of only an averaged order, can be observed. Electron microscopy is therefore uniquely suited for the examination of such disordered materials, and for the detection of disorder that may not be apparent in X-ray diffraction.

A number of examples of the use of lattice fringe imaging were contained in the first Report in this series.[1] Since then a number of papers have been published which contain information obtained largely from this technique alone. Allpress,[13] in a study of WO_3 which had been doped with up to 10 mole percent Nb_2O_5, has observed disordered CS planes and numerous structural complexities in material which had proved to be difficult to study from the point of view of X-ray diffraction.

The results may be summarized in the following way. If the amount of Nb_2O_5 doped into WO_3 is fairly low, for example $Nb_2O_5,38WO_3$, the structure consists of fairly well ordered (210) CS planes. The bulk material is fairly close to the composition $(W,Nb)O_{2.975}$, and has an approximate formula $(W,Nb)_{40}O_{119}$. Increasing the niobium content causes the CS planes to change from (210) to (410) at compositions near to $Nb_2O_5,18WO_3$, that is $(W,Nb)O_{2.95}$ or $(W,Nb)_{60}O_{177}$, and ultimately to (100) when the niobium content is increased to $Nb_2O_5,12WO_3$. This has a composition near to $(W,Nb)_{14}O_{41}$, *i.e.* $(W,Nb)O_{2.929}$. In the intermediate compositions (510) CS planes have been observed. It is likely that careful control of annealing conditions and niobium content will yield other CS plane indices. Wavy CS planes have also been observed in these materials and are shown in Figure 4. These demonstrate in a striking fashion the power of lattice fringe imaging to reveal the underlying structure of disordered CS planes. The general pattern of these results has also been confirmed by Bursill and Hyde.[14]

It is interesting to note that at very low niobium contents, up to *ca.* 1 mol % Nb_2O_5, no CS planes are observed. It would seem likely that the Nb^{5+} ions will prefer an octahedral environment in the WO_3 lattice. Such a defect can be created by the insertion of an Nb^{5+} and an oxygen atom into the structure and this has been suggested by Gadó and Magnéli.[15] Such a defect can be considered as CS plane nucleus. If these link together, more extended CS plane segments will form. The real mechanism of formation of the CS phases is likely to be unrelated to this concept, but nevertheless the structural model is attractive as it avoids the necessity of introducing point defects into the system.

Rather less work has been done on the Ta_2O_5,WO_3 ternary oxides, although Bursill and Hyde[14] have done some preliminary electron-microscope studies. They found that CS phases are not so readily formed in this system as in the niobium–tungsten oxides, and only (310) CS planes were found.

The reason for these changes may possibly be associated with the different tendencies of W, Nb, or Ta atoms to form metal–metal bonds across shared octahedron edges. Blasse[16] has suggested a similar reason to account for

[13] J. G. Allpress, *J. Solid State Chem.*, 1972, **4**, 173.
[14] L. A. Bursill and B. G. Hyde, *J. Solid State Chem.*, 1972, **4**, 430.
[15] P. Gadó and A. Magnéli, *Materials Res. Bull.*, 1966, **1**, 33
[16] G. Blasse, *J. Inorg. Nuclear Chem.*, 1964, **26**, 1191.

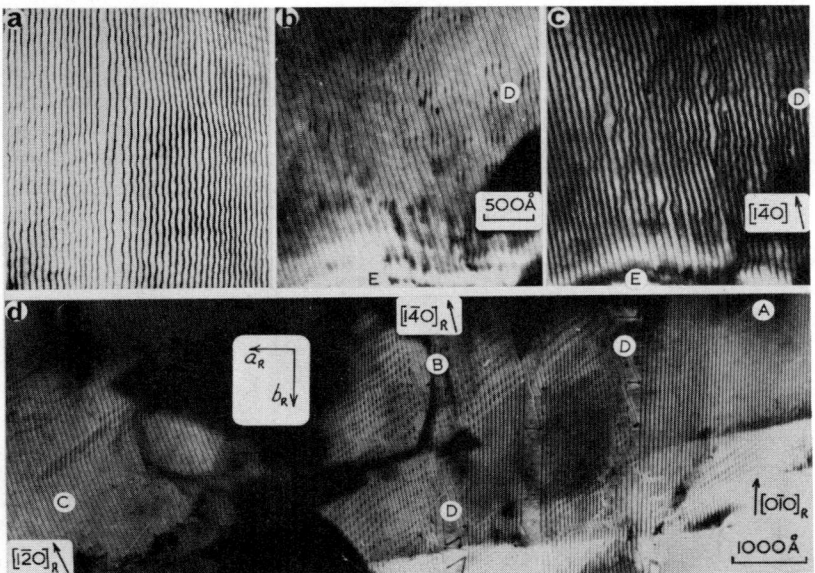

Figure 4 *Electron micrographs of fragments of crystals with a nominal composition $Nb_2O:11WO_3$. The indices and directions refer to the ReO_3-type subcell which has the same orientation in all micrographs*
(Reproduced by permission from *J. Solid State Chem.*, 1972, **4**, 173)

structural differences between other mixed-metal oxides containing these M^{5+} ions. Any such model implies, however, that the ions doped into WO_3, Nb^{5+} or Ta^{5+}, are not arranged at random, but are preferentially distributed in the CS plane itself. No careful studies have been carried out to determine to what extent this is true. Indeed, such experiments will be difficult in view of the disorder normally found in the CS phases, and the fact that the X-ray scattering factors of the ions involved are all rather similar. Allpress,[13] however, has suggested that this may be so in the Nb_2O_5,xWO_3 oxides, basing his statement upon a comparison of computed and observed $h00$ structure factors for $Nb_2O_5,12WO_3$. The neutron-scattering lengths of the atoms differ more widely and, within the block structures $Ti_2Nb_{10}O_{29}$ and $TiNb_2O_7$ neutron diffraction has revealed a segregation of the two cation types, with Ti^{4+} not being distributed at random (see Section 4).

The nature of the structures found in the binary WO_{3-x} oxides has also been determined largely by lattice-image studies. The first Report in this series[1] suggested that the earliest stages in the reduction of WO_3 appeared to be the formation of (102) CS planes. These were generally isolated from one another, taking the form of Wadsley defects. However, it was mentioned that a possible precursor of this was the formation of CS planes on (100). Bursill and Hyde[14] have confirmed that (102) CS planes are normally observed in WO_{3-x} and the evidence indicated that the presence of (100) planar faults in earlier studies[17] is not associated merely with oxygen loss in WO_{3-x}.

Further reduction in the binary tungsten–oxygen system is accompanied by extensive CS plane formation. Large WO_3 crystals lose oxygen readily. At temperatures below 900 °C the bulk structure initially contains isolated (102) CS planes. As reduction proceeds these CS planes aggregate to form quasi-ordered arrays in the WO_3 matrix, sometimes with considerable disorder. Figure 5 shows a typical fragment. The homologous series of

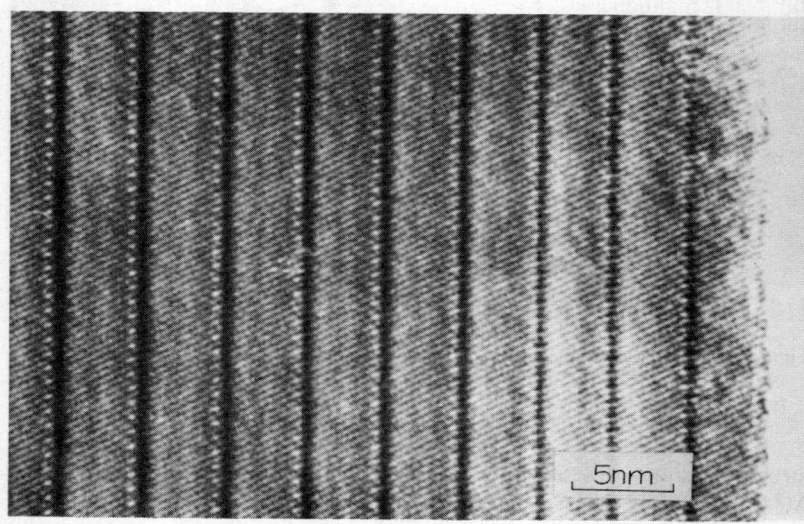

Figure 5 *Well ordered* (102) *CS planes (coarse fringes) in reduced tungsten trioxide. The crystal has been tilted a few degrees from a symmetrical (h0l) section to image the h00 planes (fine fringes). Note that the contrast in the CS plane changes as the crystal thickness increases*

oxides W_nO_{3n-1} is the ultimate result of this process. The range of n values seems to fall as low as ten or twelve, with a composition of approximately $WO_{2.90}$—$WO_{2.92}$. The (103) CS phases have not been observed to form within the bulk at these temperatures but appear as needles on the surface of the bulk crystals and at other places within reaction tubes. These most likely grow *via* a vapour-phase transport mechanism and are usually rather better ordered than the (102) CS planes. Figure 6 shows a well-ordered flake.

Bursill and Hyde[14] found fairly well isolated (103) CS planes in melted, slightly reduced samples of WO_3, suggesting that the structure type found

---

[17] J. Spyridelis, P. Delavignette, and S. Amelinckx, *Materials Res. Bull.*, 1967, **2**, 615; S. Amelinckx and J. Van Landuyt, in 'The Chemistry of Extended Defects in Nonmetallic solids', ed. L. Eyring and M. O'Keefe, North-Holland, Amsterdam, 1970.

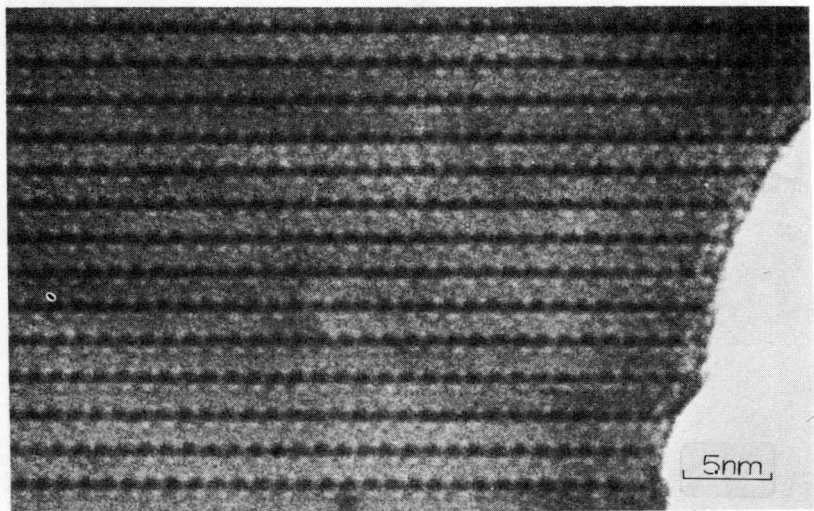

Figure 6 Well ordered (103) CS planes in reduced tungsten trioxide. The composition of the flake is close to $W_{18}O_{52}$. The groups of six edge-sharing octahedra which constitute the CS plane are resolved as darker contrast

may depend upon temperature. Recent experiments[18] have confirmed this speculation. Although sample preparation is more difficult at higher temperatures owing to the favoured formation of needle crystals from the vapour, selection of reduced fragments from large WO_3 crystals that have been carefully separated from needles shows that (103) CS planes can be found in the bulk. The conditions of formation depend upon both the temperature and the composition of the crystal. Sufficient data have not yet been accumulated to delineate this phase boundary carefully, but its position is clearly of great interest.

The oxides based upon the rutile structure have largely been understood by the use of electron microscopy. This is particularly so in the range of compositions close to MO_2 and in materials doped with altervalent cations. The study of Gibb and Anderson[19] on the $(Ga,Ti)O_{2-x}$ and $(Fe,Ti)O_{2-x}$ systems is typical of the power of CS plane imaging. In this work boundaries on (210) planes were found. In rather low concentrations of dopant, these were disordered, but in the $Ga_2O_3, xTiO_2$ oxides, fairly well ordered arrays of these planar faults were observed (Figure 7). These faults, when ordered, generate homologous series of compounds. In the case of the Ga_2O_3, TiO_2 oxides, these are of the formula $Ga_4Ti_{n-4}O_{2n-2}$. These faults are not CS planes in the normal sense, but do provide a mechanism for lowering the

[18] M. Sundberg and R. J. D. Tilley, *J. Solid State Chem.*, in the press.
[19] R. M. Gibb and J. S. Anderson, *J. Solid State Chem.*, 1972, **5**, 212.

Figure 7 Disordered faults on (210) planes in TiO_2 doped with ca.12 atom % $GaO_{1.5}$
(Reproduced by permission from J. Solid State Chem., 1972, **5**, 212)

anion to cation ratio in the material by a rather similar mechanism to the CS planes, that is, the introduction of extra 'interstitial' metal ions at the planar boundaries. This aspect of the CS phases is discussed in more detail in Section 3.

These results show that within the CS phases disorder and faulting is the rule rather than the exception. CS planes with variable spacing constitute coherent intergrowths of members of homologous series of compounds, and stoicheiometry is accounted for in this way rather than in terms of classical point defects. CS planes are also frequently found to end in the bulk. These terminate in a dislocation which provides a high-energy fault from which the CS plane can either grow or shrink. The terminating CS plane also provides another local region of stoicheiometric variability.

These features are also present in the block structures, which can be described (see Section 4) in terms of two nearly perpendicular sets of CS planes, which impose some constraints upon the flexibility of the system. Nevertheless, a very wide range of faulting has been found, often associated with the coherent intergrowth of two or more different block structures.

A number of recent papers show examples of such features in block structures. Hutchison and Lincoln[20] have examples of linear defects and anomalous block sizes in crystals prepared by doping Nb_2O_5 with MgF_2. An example of these features is shown in Figure 8 together with an interpretation of the contrast.

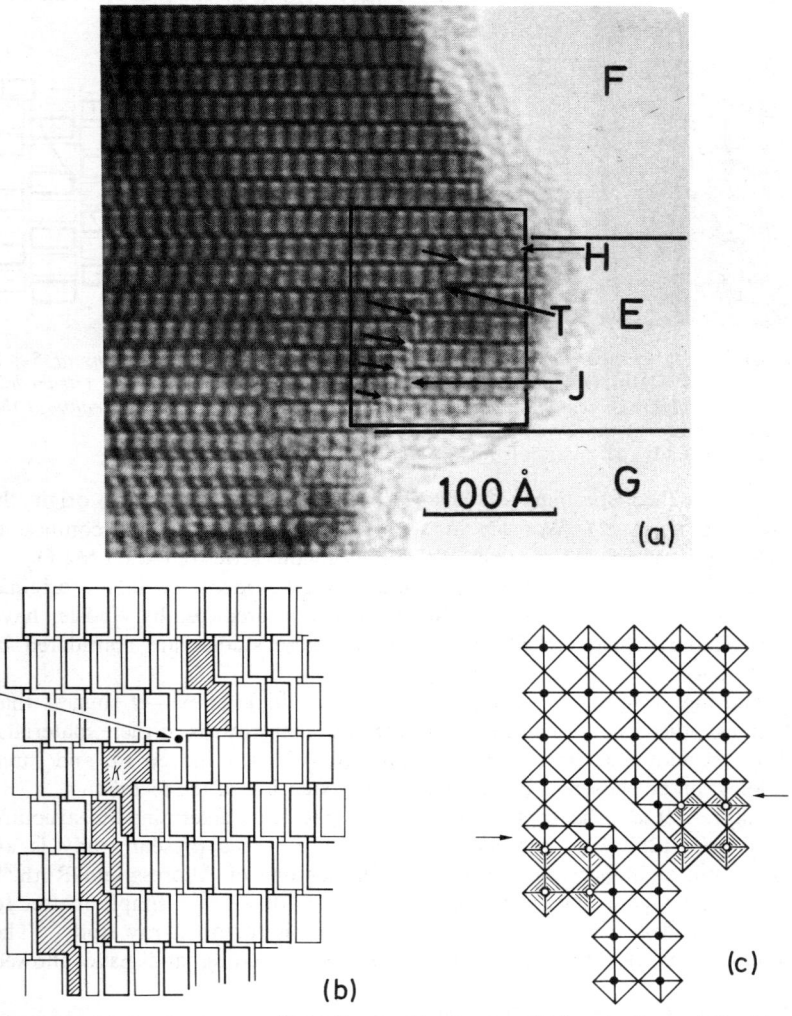

Figure 8 (a) *Lattice image of* $MgNb_{14}O_{35}F_2$ *showing* (001) *twinning and Wadsley defects*; (b) *Analysis of the structure within the enclosed area of* (a) *showing overlapping blocks and an anomalous block K*; (c) *Detailed structure of block K*
[Reproduced by permission from *Phys. Stat. Sol. (A)*, 1973, **17**, 169]

[20] J. L. Hutchison and F. J. Lincoln, *Phys. Stat. Sol. (A)*, 1973, **17**, 169.

In a remarkable series of papers, Iijima[21-25] has published high-resolution micrographs of mixed Ti–Nb oxides and H-Nb$_2$O$_5$ in which the component octahedra of the block structures are revealed. These provide conclusive proof that direct resolution of the structure of these materials is possible under the correct experimental conditions. Figure 9 shows an example from a study

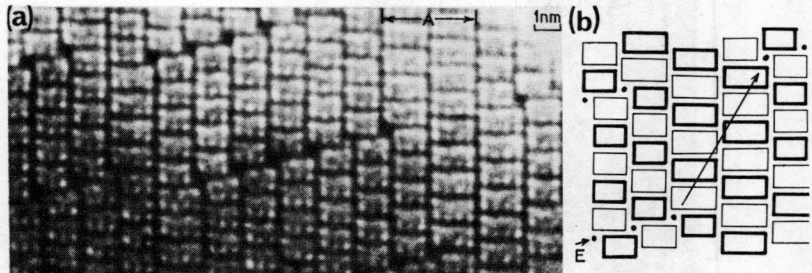

Figure 9 (a) *Lattice image showing the intergrowth of narrow domains of* 5 × 3 *blocks of* TiNb$_{14}$O$_{37}$ *and the displacement of planes* (E) *containing tetrahedral sites which are indicated by dark blobs of contrast*; (b) *Model of the region at the right end of* (a)
(Reproduced by permission from *J. Solid State Chem.*, 1973, **7**, 94)

of a quenched specimen of TiO$_2$,7Nb$_2$O$_5$. This material was originally prepared by A. D. Wadsley in an attempt to synthesize the compound TiNb$_{14}$O$_{37}$, which is a member of the homologous series of oxides M$_{3n}$O$_{8n-3}$ in which $n=5$. Although this compound was not identified in the original X-ray study, domains of the TiNb$_{14}$O$_{37}$ structure predicted by Wadsley have been found. The further implications of these studies are considered in Section 4.

The high-resolution micrographs illustrated in Figures 4–9 suggest that besides indicating the nature of faulting and intergrowths in these materials, direct structure determinations are possible. This is indeed so, and three studies can be chosen as illustrative of this extension of the technique.

Allpress *et al.*[24] have used direct lattice imaging to determine the structure of β-ZrO$_2$,12Nb$_2$O$_5$, another block structure. This present study is an extension, at higher resolution, of the earlier one of Allpress and Roth.[26] A lattice image is shown in Figure 10. From this it is a simple matter to derive an idealized structure to use as the basis for an X-ray study. The initial X-ray results confirm the postulated structure, and have allowed

[21] S. Iijima, *J. Appl. Phys.*, 1971, **42**, 5891.
[22] J. M. Cowley and S. Iijima, *Z. Naturforsch.*, 1972, **27a**, 445.
[23] S. Iijima, *Acta Cryst.*, 1973, **A29**, 18.
[24] J. G. Allpress, S. Iijima, R. S. Roth, and N. C. Stephenson, *J. Solid State Chem.*, 1973, **7**, 89.
[25] S. Iijima and J. G. Allpress, *J. Solid State Chem.*, 1973, **7**, 94.
[26] J. G. Allpress and R. S. Roth, *J. Solid State Chem.*, 1970, **2**, 366.

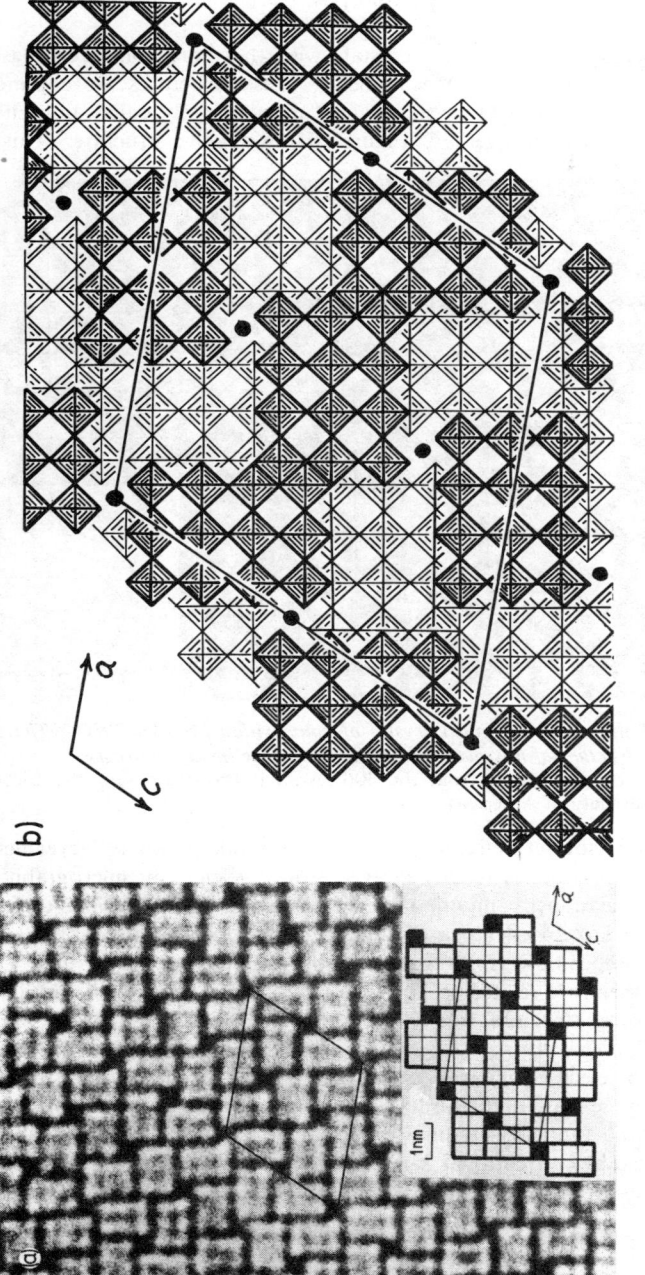

Figure 10 (a) Lattice image from a thin fragment of $ZrO_2,12Nb_2O_5$ viewed down the b axis. The inset shows a schematic representation of the contrast. (b) Idealized model of the structure, derived from (a) (Reproduced by permission from *J. Solid State Chem.*, 1973, **7**, 89)

idealized structures of other polymorphs, particularly the γ-form, to be postulated.

In another study by Iijima[27] crystals with the more complex tungsten-bronze tunnel structure were examined. These structures are commonly formed in the Nb_2O_5,WO_3 system when approximately equal amounts of Nb_2O_5 and WO_3 react together. Figure 11 shows such a structure. This is a

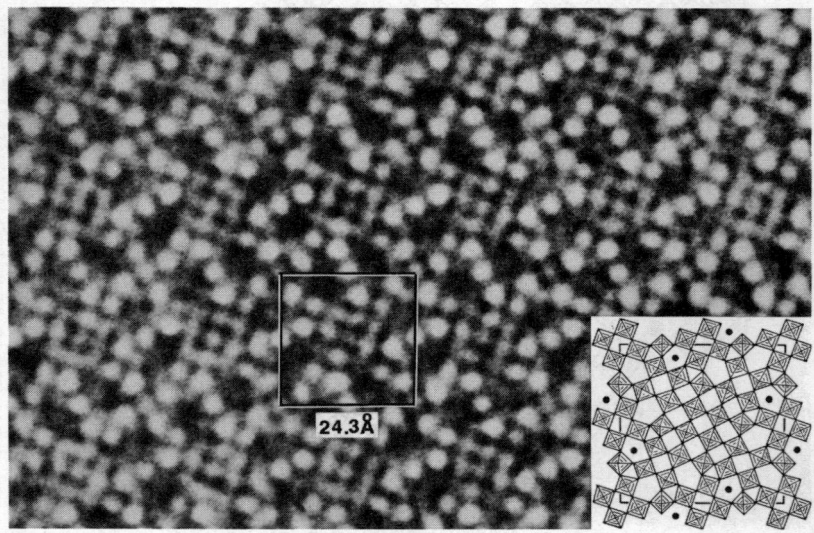

Figure 11 *Lattice image from a crystal of composition* $2Nb_2O_5,7WO_3$. *The inset shows the structure of this material, derived from the image contrast*
(Reproduced by permission from the 30th Annual Proceedings of the Electron Microscopy Society of America)

new structure, not yet determined by X-ray crystallography. Nevertheless, an idealized structure is readily derived from the electron micrograph. It is seen to be a tunnel compound related to the tetragonal bronze type but with a rather large 4×4 block of MO_6 octahedra at the centre of the unit cell. Its composition is estimated to correspond to $Nb_4W_7O_{31}$. The material is therefore related to both the block structures and the tetragonal tungsten-bronze structure, and is seen to be rather similar to the binary oxide $W_{18}O_{49}$.

The use of electron microscopy to derive idealized models of structures will probably increase in the future. This is likely to be so particularly in the CS phases and tunnel compounds which have large unit cells and are not easily handled by conventional X-ray techniques. The provision of a reliable idealized model reduces the complexity of the X-ray investigation

[27] S. Iijima, 30th Annual Proceedings of the Electron Microscopy Society of America, ed. C. J. Arceneaux, Los Angeles, 1972.

considerably. The technique is also of value for identifying small domains of phases which are not readily prepared in the form of homogeneous crystals large enough for X-ray determination. Further examples of this use of lattice imaging will be found in other sections of this Report.

Electron Microscopy and Lattice Imaging in the Study of Reactions.—The detailed observations in the preceding section can also be used as a basis from which to derive possible reaction mechanisms for the formation of CS phases. In the past this has usually taken the form of deducing the mode of CS formation in the parent oxides by an analysis of the configuration of CS planes in randomly selected crystal fragments. This approach was outlined in the previous Report.[1] A more recent paper, by Van Landuyt and Amelinckx,[28] is rather similar. From observations of thin films of TiO_2 they have suggested a mechanism of CS plane growth in which CS planes move into the foil as a pair, in a hairpin arrangement rather than separately. This mechanism may well hold in certain circumstances, but it seems unlikely that it is the only mechanism that will be applicable in the formation of CS phases. The variety of chemical compounds existing as CS phases and the widely differing atom mobilities found reinforce this concept. It therefore seems unlikely that one mechanism will be unique for all conditions of formation and all phases.

The model which appears to have the most general validity for the growth of CS planes into a bulk matrix is that described by Anderson and Hyde[29] although some minor modifications of this dislocation model have been suggested. It has been examined in greatest detail by Bursill and Hyde[30] who have considered the generation of CS planes in rutile. Using a ball model of the (100) plane of the rutile structure (idealized to give true hexagonal close packing of oxygen), they have 'grown in' CS planes by successive rearrangements of atoms. A sequence of time-lapse photographs taken with a cine camera allowed dynamic photographs of various possible reaction mechanisms to be obtained. In this way they have considered both the production of CS planes in rutile and the interconversion of various homologues one to another. These results have suggested a modification of the dislocation model in which an antiphase boundary is also involved in the movement of CS planes into rutile crystals. In this, the CS plane essentially nucleates at the surface and runs into the crystal along a path defined by an antiphase boundary.

These results do not rely to such an extent upon the high-resolution electron microscopy described in the previous sections. However, if such micrographs could be taken showing dynamical changes in the CS planes direct information upon mechanism could be found. A few such examples

---

[28] J. Van Landuyt and S. Amelinckx, *J. Solid State Chem.*, 1973, **6**, 222.
[29] J. S. Anderson and B. G. Hyde, *J. Phys. and Chem. Solids*, 1967, **28**. 1393.
[30] L. A. Bursill and B. G. Hyde, ' Progress in Solid State Chem.', ed, H. A. Reiss and J. O. McCaldin, Pergamon, Oxford, 1972, Vol. 7.

can be quoted here, in which changes take place naturally in the microscope under the combined influence of the electron beam and the surrounding vacuum of the electron-microscope column.

Iijima[31] has observed defects at this very high resolution while observing reduced crystals of $Nb_{22}O_{54}$. Figure 12 shows anomalous contrast at some block junctions. This contrast, which is attributed to tetrahedral atoms, gradually changes during observation. A model can readily be proposed from an examination of the micrographs, in which the tetrahedral Nb atoms are somewhat displaced to form perfect $Nb_{12}O_{29}$. Iijima has regarded this as a point-defect rearrangement, but in terms of lattice resolution it is possible that a chain of tetrahedral atoms exists in the matrix. The interpretation of lattice-image contrast would make this rather difficult to decide merely from inspection. If this latter supposition is correct, it then becomes quite closely related to the chains of atoms located between blocks which Anderson *et al.*[32] have characterized in the non-stoicheiometric compound $GeO_2,9Nb_2O_5$ (see Section 4).

The growth of (103) CS planes in a flake of a tungsten oxide of the W_nO_{3n-2} series under the action of the electron beam has been observed. The CS planes advance by extending along their length in accordance with the dislocation mechanism of Anderson and Hyde (Figure 13). However, a more complex rearrangement can also be seen where a loop of CS plane straightens somewhat and begins to grow parallel to the existing set. During the reaction two processes are seen to occur, the expansion of CS planes to the crystal edge, and some lateral movements to improve the ordering in the sample.

A final example of the use of electron microscopy as an adjunct to the study of reactions is provided by the studies of Anderson *et al.*[33,34] on the chemistry of niobium oxides. In a thermogravimetric study of the reduction of H-Nb_2O_5, the structural analysis was carried out by electron microscopy and diffraction, as X-ray techniques were not sufficiently sensitive to follow the behaviour. It was found that the course of the reaction, which traversed the composition range $NbO_{2.50}$—$NbO_{2.42}$, was structurally determined by the geometry of the starting phase. This was Nb_2O_5, with a block structure of (4 × 3) and (5 × 3) blocks. The first reaction to take place is one in which the (5 × 3) columns of ReO_3 structure in H-Nb_2O_5 are rearranged to (4 × 3) columns. At temperatures above 1100 °C the regions which contain only (4 × 3) blocks are able to intergrow with H-Nb_2O_5 to form the 1 : 1 intergrowth phase $Nb_{53}O_{132}$ before complete conversion into the more reduced material $Nb_{25}O_{62}$, which consists solely of (4 × 3) blocks. At lower temperatures the intergrowth is not formed, presumably because ordering the

---

[31] S. Iijima, 31st Annual Proceedings of the Electron Microscopy Society of America, ed. C. J. Arceneaux, New Orleans, 1973.
[32] J. S. Anderson, D. J. M. Bevan, J. M. Browne, A. K. Cheetham, R. Von Dreele, J. L. Hutchison, F. J. Lincoln, and J. Straehle, *Nature*, 1973, **243**, 81.
[33] J. S. Anderson and K. M. Nimmo, *J. C S Dalton*, 1972, 2328.
[34] J. M. Browne, J. L. Hutchison, and J. S. Anderson, 'Reactivity of Solids', 7th International Symposium on Reactivity of Solids, Chapman and Hall, London, 1972.

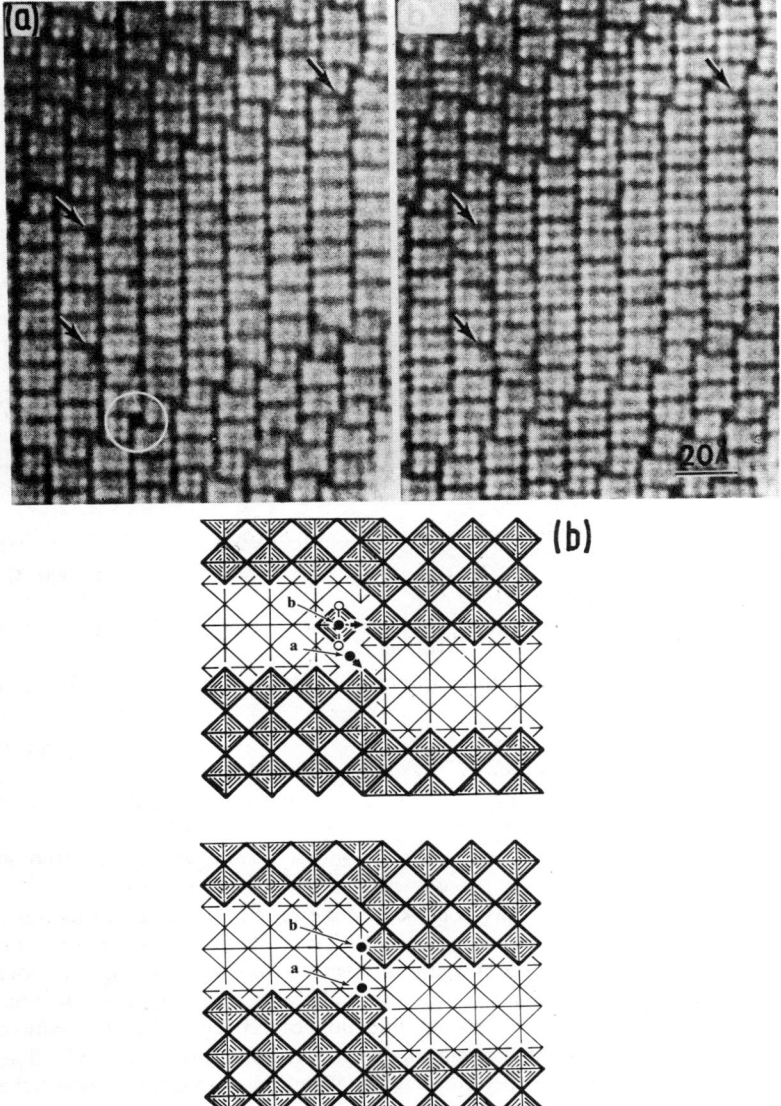

Figure 12 (a) *Successive lattice images showing defects which disappeared during observation.* (b) *A proposed model of the defect in* (a) *and its annihilation*
(Reproduced by permission from the 31st Annual Proceedings of the Electron Microscopy Society of America)

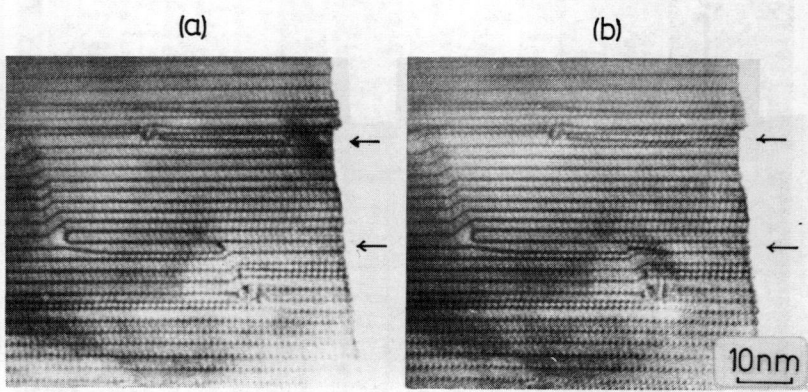

Figure 13 *Lattice images of* (103) *CS planes in reduced tungsten trioxide. The CS plane structure has altered during observation from that in* (a) *to* (b)

intergrowth has a higher activation energy. The process can then be written

$$\underset{(5 \times 3)_\infty + (4 \times 3)_1}{Nb_{28}O_{70}} \longrightarrow \underset{(4 \times 3)_2}{Nb_{25}O_{62}} \quad \text{at 1000 °C}$$

$$Nb_{28}O_{70} \longrightarrow \underset{1:1 \text{ intergrowth}}{Nb_{53}O_{132}} \longrightarrow Nb_{25}O_{62} \quad \text{at 1100 °C}$$

Further reduction proceeds by an analogous process, and can be formally written as

$$\underset{(4 \times 3)_2}{Nb_{25}O_{62}} \longrightarrow \underset{(4 \times 3)_\infty}{Nb_{12}O_{29}} \quad \text{at 1000 °C}$$

$$\underset{(4 \times 3)_2}{Nb_{25}O_{62}} \longrightarrow \underset{1:1 \text{ intergrowth}}{Nb_{47}O_{116}} \longrightarrow \underset{(4 \times 3)_\infty + (3 \times 3)_1}{Nb_{22}O_{54}} \longrightarrow$$

$$\underset{(4 \times 3)_\infty}{Nb_{12}O_{29}} \quad \text{at 1100 °C}$$

The reaction takes place so as to preserve coherence in the block structures.

When the reverse process is considered, that is, the oxidation of $Nb_{12}O_{29}$ to Nb_2O_5, electron microscopy again is able to follow the reaction in a more satisfactory way than *X*-ray techniques. It was found that monoclinic $Nb_{12}O_{29}$ transformed to a new modification of Nb_2O_5, $Nb_{10}O_{25}$, with a (3×3) block structure, at the remarkably low temperature of 400 °C. The reaction proceeds *via* a superlattice phase which is a necessary part of the reorganization process. The lattice images reveal quite clearly the structure of the final Nb_2O_5 phase. It is apparent that this particular modification of Nb_2O_5 is produced because of the structural constraints of the parent phase, the low temperature of reaction making severe structural rearrangements unattractive. At higher temperatures it is conceivable that a different structural path will be followed, although this will not necessarily be the case.

3 Rotating CS Planes

CS Structures in the Titanium and Titanium–Chromium Oxides.—The earlier electron-microscope results indicated that CS planes displayed considerable disorder, with variable spacings between the CS planes. At this time it was believed that this variable spacing was the principal mode by which these materials accommodated variable composition. In the titanium oxides two series of compounds fell into this pattern. In the composition range $TiO_{1.75}$—$TiO_{1.90}$ a homologous series existed based on (121) CS planes, while between $TiO_{1.93}$ and TiO_2, (132) CS planes were found; these latter were ordered at the lower end of this range, but disordered near to TiO_2. The major problem to be solved was the structure of the region $TiO_{1.90}$—$TiO_{1.93}$, but it was believed that in this phase interval, $Ti_{10}O_{19}$, of the (121) series, coexisted with $Ti_{16}O_{31}$ of the (132) series. The spacings of the CS planes in these two phases are the same, although their orientation is different.

Bursill, Hyde, and Philp[35] have clarified the nature of this region and shown that change in stoicheiometry is accommodated in a far more subtle fashion. In the binary system electron-diffraction patterns suggested the existence of CS planes intermediate between (121) and (132), but these samples were heavily disordered and a study of the ternary system $(Ti,Cr)O_x$ was necessary to obtain a complete understanding of these intermediate CS phases.[35, 36]

It transpires that the CS plane orientation swings progressively from one orientation to the other in the phase range between the (121) and (132) ordered phases; this is more easily discernible in the ternary oxides than in the binary oxides. The structural geometry of these boundaries, which is considered in detail in the following section, shows how the CS planes in this intermediate range have indices (*hkl*) that can be resolved into segments of (121) CS plane alternating with segments of an antiphase boundary on (011). Each CS plane can thus be considered as an ordered intergrowth of units of (121) boundary and units of (011) boundary. The indices found vary from (132) through (253), (374), and (495) to (121). In the (253) series, most of the possible phases $(Cr,Ti)_nO_{2n-2}$ with n falling between 28 and 38 were observed, but other phases were identified, and in many cases departures from the (121) orientation can be so slight as to be discernible only by careful examination of diffraction patterns from crystals oriented to make the [1$\bar{1}$1] zone clearly visible.

Attention may be drawn to some important features of these intermediate phases. The first of these is that the orientation (*hkl*) of the CS planes in any crystal is usually well defined, although the indices may be high. A change in composition can then be accommodated not only by a change of spacing between the CS planes, but also by a change in orientation, to produce a

[35] L. A. Bursill, B. G. Hyde, and D. K. Philp, *Phil. Mag.*, 1971, **23**, 1501.
[36] R. M. Gibb and J. S. Anderson, *J. Solid State Chem.*, 1972, **4**, 379.

new ordered phase. All the CS planes in this region have roughly the same spacing between CS planes: 1.61 nm between (132) CS planes in $Ti_{16}O_{31}$ and 1.60 nm for $Ti_{10}O_{19}$, of the (121) series. The spacing is more variable in the $(Ti,Cr)O_x$ oxides; for example, the (253) series covers a fairly wide composition range. In these phases, the spacing varies from 1.74 nm for $n = 28$ to 3.03 nm for $n = 48$.

There is also some uncertainty about the microstructure of these high-index CS planes. Bursill and Hyde have taken lattice fringe images of high-index CS planes which are clearly segmented into varying lengths of two components. For example, they have observed CS planes consisting of alternating segments of (132) and (253) orientation, whereas the diffraction pattern indicated a perfect CS phase based upon (385) CS planes. This clearly reveals both the averaging effect of the diffraction pattern and the power of lattice resolution. On some high-resolution photographs, though, the CS planes are straight. Thus the distribution of the unit segments making up these higher CS planes can be arranged either in regular sequence at the unit-cell level, or in a coarser distribution that can be likened more to a mixture.

In the binary system, the reversible behaviour found by Merritt for oxidation and reduction in the composition range $TiO_{1.9}$—$TiO_{1.93}$ and described in the earlier Report[1] can now be related to the structural changes described. Composition changes do not involve varying proportions of the two end members of the two CS series. Instead, a gradual reorientation of CS planes preserves good internal order without involving any change in the number of CS planes. This type of rearrangement is apparently structurally facile whereas the introduction or elimination of CS planes is a more difficult process which results in severe hysteresis in the partial molar free energy curve.

CS Structures in the Tungsten Oxides.—Far less work has been done on the tungsten oxides than on the titanium oxides. Experimental studies are outlined in Section 2, but it may be noted that the $(Nb,W)O_x$ system bears some analogies to the $(Ti,Cr)O_x$ system in that a continuous change of CS plane indices is found, from (103) to (001), over a relatively small change of composition. In the binary tungsten–oxygen system these changes are less easy to follow, and further studies are required before the system is as well understood as titanium oxides.

The Topology of CS Structures.—The results described above reveal a great variation in structural types amongst the CS phases already known. This complexity is more apparent than real and can be rationalized in terms of two lattice displacements, a lattice collapse and an antiphase displacement. The former is a pure CS operation and it alters the stoicheiometry of the crystal by a small amount. If, however, displacement lies parallel to the plane joining the two parts of the crystal, no change in stoicheiometry results but the cation rows in the two parts are mutually displaced. The complexity of CS plane types can now be constructed by intergrowing elements of collapse and antiphase relation in varying proportions through the operation of the same displacement vector. It may be observed that the permitted displacement vectors must result in an identity operation when they act upon the anion sublattice. This is illustrated for the two principle systems of CS, the ReO_3 group and the rutile group.

ReO_3-*related Structures.* The structural geometry of the ReO_3 phases has been discussed by Bursill and Hyde.[14]

The ReO_3 lattice, shown in Figure 14, is able to collapse on one of the equivalent ($h00$), ($0k0$), or ($00l$) planes. A conservative displacement of the lattice can also occur without any change in stoicheiometry, the simplest being along one of the {101} set of planes. The commonly observed {102} and {103} CS planes can now be resolved into segments of both the ($h00$) and (101) boundaries. The indices of a CS plane ($h0l$) are given by the equation

$$(h0l) = p(001) + q(101) = (q, 0, p+q)$$

if the CS planes lie parallel to b and the 010 zone is considered. Hence

$$(102) = 1.(001) + 1.(101)$$
$$(103) = 2.(001) + 1.(101)$$

The amount of stoicheiometric change is governed solely by the amount of the component of collapse present, that is, the integer p, and the formula of a homologous series based upon a set of ordered ($q, 0, p+q$) CS planes is given by the formula M_nO_{3n-p}. The (102) and (103) series then are related to the series formulae M_nO_{3n-1} and M_nO_{3n-2}, as has been known since their discovery.

In the last two cases the value of p changes from 1 to 2, and the number of pairs of octahedra linked in each group, $p+1$, changes from 2 to 3. This can be continued to give 4, 5, 6 ... pairs of linked octahedra. The resulting CS planes then lie on (104), (105), (10p) ... (001). These are illustrated in Figure 14.

This situation is found experimentally in the $(Nb,W)O_{3-x}$ oxides. For small niobium concentrations, the CS plane is {102} whereas for small tungsten concentrations the CS planes are on (001), although two intersecting sets are then always present and the oxides have block structures of the H-Nb_2O_5 type. At intermediate concentrations, the CS plane type changes progressively from {102} to {103} to {104} and so to one set of {001} CS planes when the niobium concentration is of the order of 10 mole per cent.[13]

Any CS plane lying between any adjacent pair of CS planes in the group above with $q = 1$ must embody some units of each of the adjacent pairs. Thus the plane of lowest index between (102) and (103) is (205), constructed by combining units of (102) and (103) structure alternately along the CS planes. Similarly the next lowest are (307) and (308), which are made up of two units of (102) plus one of (103) and two units of (103) plus one of (102) respectively as shown in Figure 15.

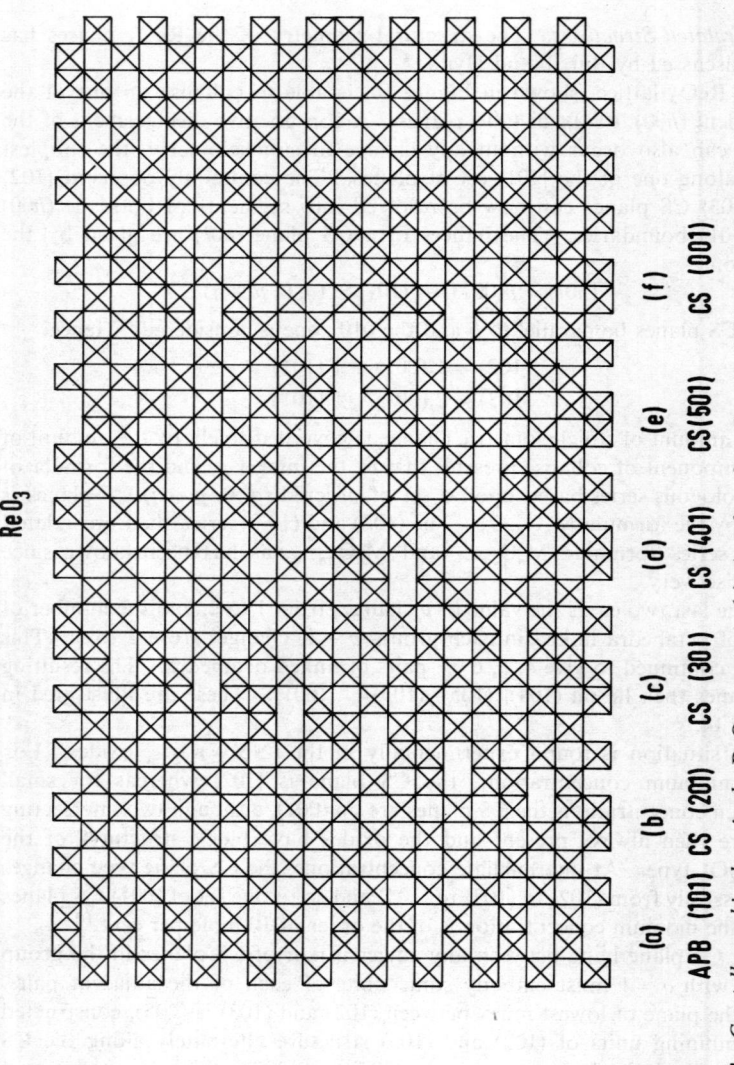

Figure 14 *Crystallographic shear in the* ReO_3 *structure.* (a) *Antiphase boundary on* (101); *crystallographic shear on* (b) (201), (c) (301), (d) (401), (e) (501), *and* (f) (001)

Crystallographic Shear and Non-stoicheiometry

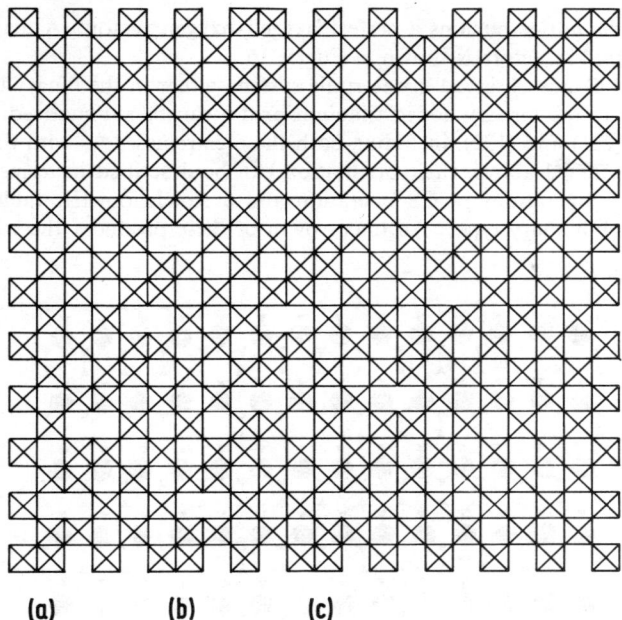

Figure 15 *Crystallographic shear in the* ReO_3 *structure on* (a) (703), (b) (502), *and* (c) (803)

In these cases with $q > 1$, the number of edge-shared pairs of octahedra in a unit of CS plane is $(p + q)$, and they are arranged in q separate groups. If the alternating groups are not ordered, the CS plane will either wander between the two end members or be resolved into zig-zag sections, depending upon the CS plane groupings.

As a result of these possibilities, almost any stoicheiometry could be accommodated within an ordered CS structure; it is also possible that a given stoicheiometry may be achieved in more than one way; thus the composition $WO_{2.91}$ can be made from either an ordered (102) or a (103) CS structure. This aspect of these phases will be considered again later.

Rutile-related Structures. The rutile structure can be regarded as derived from an MO_3 parent of the PdF_3 structure type, by the recurrent operation of a CS vector $\frac{1}{3}[112]/(100)_{PdF_3}$. If the collapse is upon the (001) planes, with a vector $\frac{1}{3}[121]$, the α-PbO_2 structure results.

Since the previous Report, Bursill and Hyde[30] have clarified the structural geometry of the rutile-based CS phases so as to remove much of the complexity of the experimental results. For this they idealize the oxygen packing in the rutile structure, from the puckered arrangement found in practice to perfect h.c.p., by flattening on $(h00)$ or $(0k0)$. If the $(h00)$ planes are flattened, they become the $(00l)$ planes of the hexagonal lattice and the CS vector $\frac{1}{2}[0\bar{1}1]$

in the rutile lattice remains a perfect oxygen–oxygen vector. A diagram of this idealized structure is shown in Figure 16.

The CS planes observed in the rutile system, either in the binary titanium oxides or in the ternary doped titanium oxides can be summarized as lying between (121) and (132), including both these extremes. As with the ReO_3 oxides discussed above, their orientation can be described in terms of two elementary operations: a CS operation and a stoicheiometric fault of the antiphase boundary type. An intergrowth of these produces the variations observed.

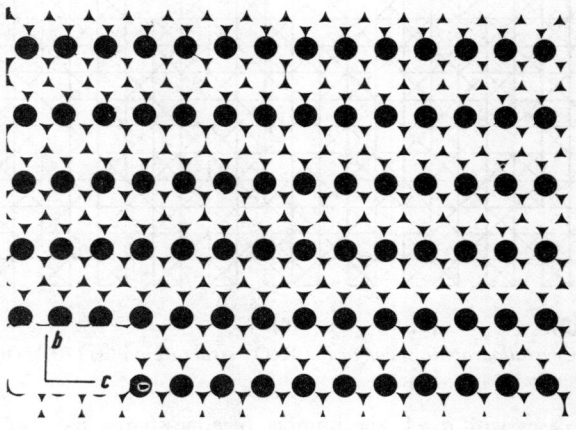

Figure 16 *A (100) layer of idealized rutile*

An antiphase boundary (APB) is produced by the vector $\frac{1}{2}[0\bar{1}1]$ operating on (011). The strings of edge-shared octahedra are stepped across the APB as can be seen from Figure 17. It is convenient to represent such a boundary of steps by the sequence

$$\ldots\ldots\ldots ///////// \ldots\ldots\ldots$$

The same fault vector, $\frac{1}{2}[0\bar{1}1]$, operating on (121) eliminates oxygen sites from the rutile structure. The structure of the (121) CS plane so formed differs from the APB described above in having an extra cation at each step along the trace of the boundary. This extra cation is not to be regarded as a point defect in any sense: it is an integral part of the structure. If one is to equate anything with a defect in the rutile lattice it should be the CS plane as a whole. Representing the CS step by the symbol ∇, which denotes the folding of the cation strings into a Z shape, the trace of the CS plane on (100) can now be represented as

$$\ldots\ldots\ldots \nabla\nabla\nabla\nabla\nabla \ldots\ldots\ldots$$

Ordered sequences of ∇ and / units can be formulated as has been done by Bursill and Hyde;[30] the number of possible ordered combinations of these

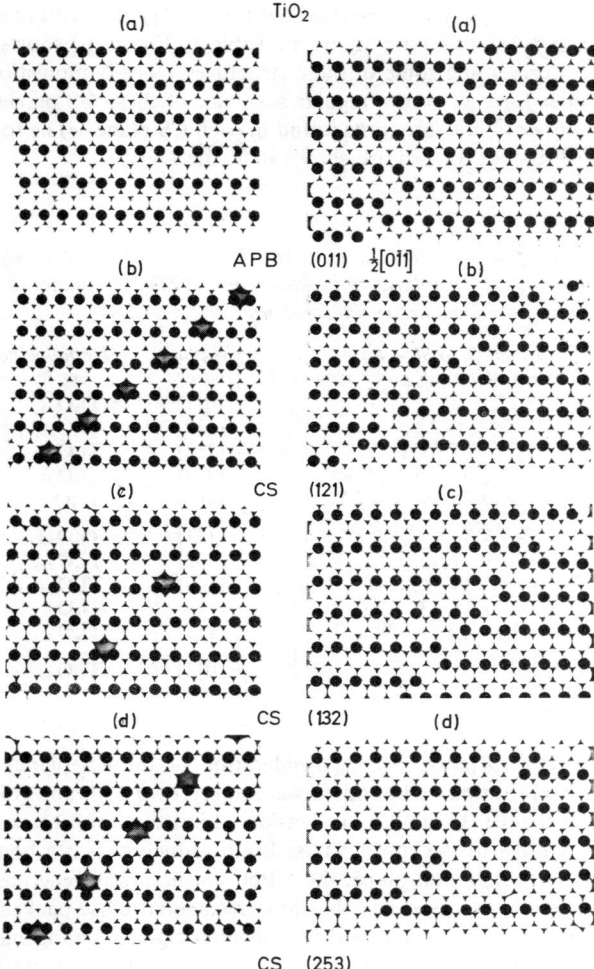

Figure 17 *Crystallographic shear in the rutile structure. The structures on the left show the pattern of vacant oxygen sites needed for subsequent collapse to produce the structures on the right:* (a) *an antiphase boundary on* (011); *crystallographic shear on* (b) (121), (c) (132), *and* (d) (253). *The displacement vector for* (a) *to* (d) *is* $\frac{1}{2}[0\bar{1}1]$

two elements is very large. Of special interest are the (121) and (132) CS planes found in the rutile system. The latter represents an ordered array of APB (/) and collapse (∇) components:

$$\ldots\ldots\ldots/\nabla/\nabla/\nabla/\nabla/\nabla/\ldots\ldots\ldots$$

The structures and the indices of the swinging CS planes lying between (121) and (132) and found in the $(Cr,Ti)O_{2-x}$ oxides are those obtained by the intergrowth of different proportions of these (121) and APB components. Some of these possibilities are shown in Table 1; Figure 17 shows idealized (100) layers representing some of these structures. The composition of each series of oxides formed from a regular sequence of these CS planes depends upon the number of collapse units found in each CS plane. The formulae of some series of oxides formed are shown in Table 1.

Table 1 *Some possible structures formed by the ordered intergrowth of elements of antiphase boundaries and (121) CS planes in titanium oxides based on the rutile structure*

Sequence of structure elements along CS plane trace	CS plane indices	Homologous series
.../ / / / / / / /...	(011)	Ti_nO_{2n}
.../∇/∇/∇/∇/∇/...	(132)	Ti_nO_{2n-1}
.../∇∇/∇∇/∇∇/∇∇/...	(253)	Ti_nO_{2n-2}
.../∇∇∇/∇∇∇/∇∇∇/∇∇∇/...	(374)	Ti_nO_{2n-3}
.../∇∇∇∇/∇∇∇∇/∇∇∇∇/...	(495)	Ti_nO_{2n-4}
...∇∇∇∇∇∇∇...	(121)	Ti_nO_{2n-1}
...∇/∇/∇/∇/∇/...	(132)	Ti_nO_{2n-1}
...∇//∇//∇//∇//∇...	(143)	Ti_nO_{2n-1}
...∇///∇///∇///∇...	(154)	Ti_nO_{2n-1}
...∇////∇////∇////∇...	(165)	Ti_nO_{2n-1}
.../ / / / / / / /...	(011)	Ti_nO_{2n}

α-PbO$_2$-*related Structures.* In a consideration of the CS planes in rutile, units of the (011) antiphase boundary were intergrown with units of the (121) CS plane. If the (011) boundary is considered in isolation it causes a step in the cations parallel to the rutile *c* axis; By introducing such a boundary on every alternate (011) anion plane, the α-PbO$_2$ structure is generated (Figure 18). The α-PbO$_2$ structure has the same hexagonal close-packed array of oxygen atoms as in rutile, but with a staggered zig-zag arrangement of the strings of edge-shared cation–anion octahedra. It can be regarded as an ordered polysynthetic rutile. If the (011) boundaries are inserted on every anion plane, a twinned orientation of rutile is found.

In the first Report, CS operations in the α-PbO$_2$ structure were considered from a geometrical point of view. At that time, no CS phases based upon this structure had been reported. Since then Grey and Reid[37] have published a fairly detailed study of phase relations in the system Fe–Cr–Ti–O. For certain compositions both pseudobrookite phases and oxides based on (121) CS in rutile were recorded, but they found a series of structures with a general formula $(Cr,Fe)_2Ti_{n-2}O_{2n-1}$ over much of the phase range. These were

[37] I. E. Grey and A. F. Reid, *J. Solid State Chem.*, 1972, **4**, 186.

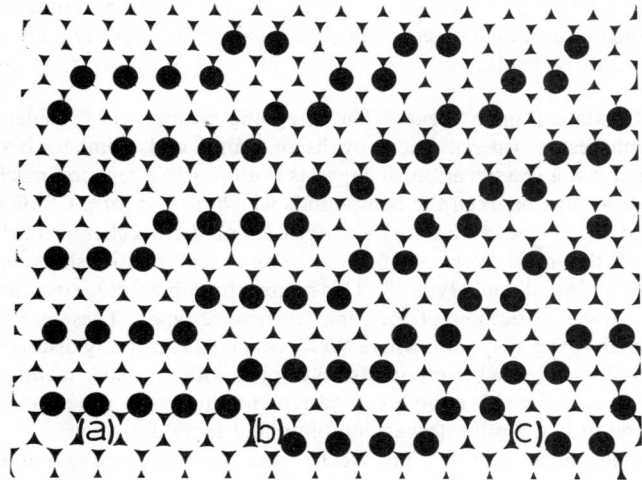

Figure 18 A (100) *layer of idealized rutile* (a), *containing* (b) *a single* (011) *antiphase boundary and converting into* (c) *a lamella of the* α-PbO$_2$ *structure*

members of a homologous series of oxides based upon CS in an α-PbO$_2$ parent.

The structures of these phases were established through consideration of the structure of V$_3$O$_5$,[38] which was found to be isostructural with(Cr,Fe)$_2$TiO$_5$, and a determination of the structure of the M$_4$O$_7$ homologue, CrFeTi$_2$O$_7$.[39] The V$_3$O$_5$ structure can, in fact, be regarded either as $n = 3$ in the (121) CS phases derived from rutile or as $n = 3$ in a series based upon (110) CS planes in an α-PbO$_2$ parent. The higher homologues $n = 4$ and 5 are also derived from the same model, with an increased spacing between the CS planes.

Besides these phases, ordered intergrowths have also been found between the M$_4$O$_7$ homologue and the members on either side, M$_3$O$_5$ and M$_5$O$_9$, and these can be written as series of oxides (M$_3$O$_5$)$_m$(M$_4$O$_7$)$_n$ and (M$_4$O$_7$)$_n$(M$_5$O$_9$)$_n$. Because of their similar structures all these intergrowth phases give very similar X-ray powder patterns. Grey and Reid have suggested that the apparent range of homogeneity shown by the ternary phase Cr$_2$Ti$_2$O$_7$ may well be explained on this basis. It would be very difficult, from powder X-ray work alone, to identify such intergrowth structures, even if they were perfectly ordered.

The α-PbO$_2$ structure is a high-pressure polymorph of rutile, and the CS planes in these homologues can be considered to stabilize this structure to atmospheric pressure. The width of the slabs which can be stabilized in this way is limited, and in these materials does not exceed that of the $n = 5$ oxide.

[38] S. Åsbrink, S. Friberg, A. Magnéli, and G. Andersson, *Acta Chem. Scand.*, 1959, **13**, 603.
[39] I. E. Grey and W. G. Mumme, *J. Solid State Chem.*, 1972, **5**, 168.

The somewhat larger zirconium ion can be made to replace titanium in these compounds to a limited extent, and such a substitution permits homologues with $n > 5$ to be made.

Rotary Crystallographic Shear.—The foregoing geometrical consideration of the CS phases as ordered intergrowths of simple end members is of more value than just a classification; it suggests both possible reaction mechanisms and relationships between the compounds which do not form CS phases and those that do. These in turn are then useful for predicting chemical behaviour.

One of the most elegant of these geometrical relationships has been described by Bursill and Hyde.[40] This relates the cubic ReO_3 structure to the complex tunnel structures of the tungsten bronze type. These are found in the tungsten–oxygen system, where WO_3, which has a slightly distorted ReO_3 structure, forms tunnel compounds on doping with an alkali metal. Similar structures exist in the complex tungsten–niobium and tungsten–tantalum oxides and in many other ternary niobium and tantalum oxides.

The relationship between the ReO_3 parent and the tetragonal tungsten bronze structure is shown in Figure 19. The basic operation required is the

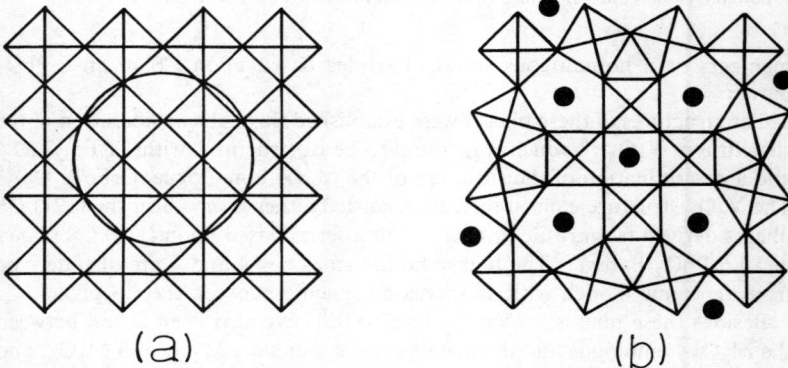

Figure 19 *The formation of the framework of the tetragonal tungsten bronze structure (b) from the ReO_3 structure (a) by rotation of the circled unit by 45°, interpolating alkali-metal atoms (●) in some tunnels*

rotation of a column of ReO_3 octahedra by 45°. In some ways this is akin to the generation of an antiphase boundary in ReO_3. In this case, however, the antiphase boundary has changed from a planar fault into a narrow cylinder, an antiphase column. This operation produces pentagonal tunnels. The relative positions of the rotation centres of the antiphase columns now yields a variety of groupings of pentagonal or, if a suitable juxtaposition is arranged, hexagonal tunnels, which can now form the frameworks both of the bronze structures already characterized and of a large number of hypothetical

---

[40] L. A. Bursill and B. G. Hyde, *Nature Phys. Sci.*, 1972, **240**, 122.

structures. The real structures differ from these idealized frameworks in that the metal–oxygen polyhedra are distorted, and the tunnels frequently contain additional cations. These are not introduced by the simple rotation, which preserves the MO_3 stoicheiometry of the parent, but may be important in stabilizing a more open skeletal structure.

There is a close relationship between the ReO_3-based CS phases and the tunnel structures. Both are formed by reduction of a hypothetical or real ReO_3 parent. The range of composition over which the CS phases are favoured depends on the material studied. For MO_{3-x} this is approximately MO_3—$MO_{2.88}$, and the CS structure formed depends upon the metal M present. The niobium oxide block structures have a range $MO_{2.70}$—$MO_{2.3}$. Both systems form tungsten bronze or related structures when these limits are exceeded (compositions MO_3—$MO_{2.5}$) and coexist with the rutile structure when the composition is below $MO_{2.3}$. The degree of reduction is not the only variable; large cations with a low charge, the alkali metals, and alkaline earth metals tend to force the ReO_3 matrix to adopt a tunnel structure over quite substantial composition ranges, whereas the smaller, more highly charged transition-metal ions are more tolerant of the CS phase structure. This may well be in part due to size effects, plus the fact that transition-metal ions in general have an energetic preference for octahedral co-ordination. The larger M^{1+} and M^{2+} ions cannot fit so readily into an almost close-packed oxygen array and force the structure type to change more drastically.

These variables are not the only ones to consider. The oxide $W_{18}O_{49}$ has a structure which can be best described as a tunnel structure, but which is derived from an ReO_3 parent by a combination of rotation and planar CS operation. The questionable valence of the tungsten cations in this structure makes it difficult to suggest that size or low valence themselves are responsible for these structural changes, and further experiments may well show more subtle causes.

As the composition of the transition-metal oxides approaches MO_2, the stable structure type becomes, in most cases, the rutile structure, into which rotation faults of the same type as in the ReO_3 matrix can be introduced. These form the basis for a series of compounds in many ways analogous to the tungsten bronzes, the hollandites, typified by the manganese oxides $A_xMn_8O_{16}$, where A can be a large uni- or bi-valent ion, Ba^{2+} being typical. The generation of these structures is shown in Figure 20.

There are close parallels between the formation of the hollandites and the tungsten bronzes. In both, cations of low valence and large radius, which cannot occupy octahedral interstices, are introduced into a fairly close-packed oxygen array. If the cations are smaller and more highly charged, they can fit in, and either form mixed rutiles or trirutiles or else develop CS phases based upon the TiO_{2-x} or $\alpha\text{-}PbO_{2-x}$ structure types. As before, simple considerations of size seem insufficient at this stage to account for all the observations and more subtle forces may be responsible.

The rutile structure type can also be related topologically to the fluorite

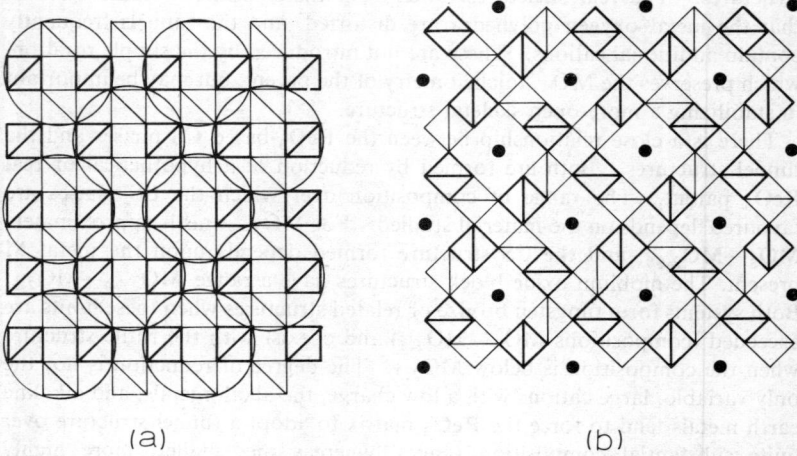

Figure 20 *The formation of the hollandite structure* (b) *from the rutile structure* (a) *by rotation of the units circled by* 45° *and interpolating metal atoms* (●) *in the tunnels. The* [MO_6] *octahedra are idealized to squares in this projection*

structures.[41] CS has not been found hitherto in the fluorite-related phases, but has been discussed by Hyde[42] as being relevant to the chemistry and possible CS transformations of UO_2 and related phases.

4 Block Structures

The Topology of Block Structures.—It is possible to extend the foregoing discussion of the structure and rotation of crystallographic shear planes to give a rationale for the block structures found in the niobium oxides and related compounds. Their characteristic is that they are built up from rectangular columns of ReO_3 structure, measuring $m \times n$ octahedra in cross-section and infinite in length. The octahedrally co-ordinated cations are at different levels ($z = 0$ and $z = \frac{1}{2}$) in adjacent columns, so that they share octahedron edges at the interface between columns. These interfaces thus have the structure of a short segment of {100} CS plane. A perfectly ordered structure can be adjusted to subtle variations of composition by changes in the cross-section and linkage of the columns,[1] which change the number of anion sites eliminated by octahedron edge sharing.

This extensive set of elegant, but at first sight complex, structures can be related to the parent ReO_3 structure type by the simultaneous, recurrent operation of two nearly orthogonal CS operations: $(\frac{1}{2}a)[011]/(h0l)$ with $l > h$ and $(\frac{1}{2}a)[\bar{1}10]/(h'0l')$ with $h' > l'$. In defining these CS operations, the

---

[41] B. G. Hyde, L. A. Bursill, M. A. O'Keeffe, and S. Andersson, *Nature Phys. Sci.*, 1972, **237**, 35.
[42] B. G. Hyde, *Acta Cryst.*, 1971, **A27**, 617.

displacement vector and the crystallographic shear plane are stated, here and subsequently, in terms of the uncollapsed ReO_3 structure. Except for {101}, any oblique ($h0l$) plane intersecting the square (010) cation lattice of the ReO_3 structure necessarily has its trace divided into segments; for $l > h$, these consist of steps with treads parallel to (001) and risers, of unit height, parallel to (100). Writing $l/h = I + R/h$, where I is an integer and $0 \leqslant R < h$, the segments will consist of R steps of length $(I + 1)$ and $(h - R)$ steps of length I. Thus all CS plane orientations between (104) and (105) generate segments of (001) structure, four and five octahedra long; for orientations between (103) and (104), the segments are three and four octahedra long. Similarly, a second shear operation on an orientation close to (100) generates a stepped trace with all interfaces along (100). Recurrent crystallographic shear in both directions automatically cuts the parent ReO_3 structure into rectangular columns, as shown for the hypothetical generation of the H-Nb_2O_5 structure in Figure 21. Structures that have one CS plane direction in common can intergrow coherently on that plane; the second CS plane is kinked as it passes from one structure to the other. If, now, with one CS plane held constant, the other rotates, with concomitant change of composition, the relative proportions of the two types of {100} segment are altered, thereby changing the cross-sections of the blocks. This is illustrated in Figure 22 for structures with one common CS operation on $(601)_{ReO_3}$; a second CS operation on $(10\bar{4})_{ReO_3}$ generates $Nb_{25}O_{62}$ or, on $(20\bar{9})_{ReO_3}$, H-Nb_2O_5. If the CS surface remained planar, rotation between these two directions would give rise to regular intergrowth structures, with ordered, but in general complex, successions of steps in the CS plane. The simpler regular intergrowth sequences are known; the more complex sequences cannot be realized experimentally in a well ordered form, but have been found by electron microscopy as disordered intergrowths with kinked or wavy shear planes, corresponding to the out-of-equilibrium structures that have been observed in the simpler CS phases. Table 2 shows the relation between the parent ReO_3 structure and some block structure compounds; from the CS operator description follows the ability of different structures to generate intergrowth phases and to accommodate Wadsley defects through the kinking of CS planes.

The spacing between regularly recurrent CS planes prescribes the sizes of the columns enclosed between them. In the binary niobium oxides, block sizes smaller than (3 × 3) or larger than (3 × 5) octahedra—corresponding to normal distances 1.0—1.7 nm between CS planes—are not found, except as faults (*i.e.* as metastable configurations). Larger blocks, up to (5 × 5) and (3 × 6) octahedra, are found in the ternary WO_3–Nb_2O_5 and niobium oxyfluoride phases.[1] These limitations can be broadly understood in terms of Pauling's rules. The central octahedra in a block have the environment of the ReO_3 structure, little perturbed; an octahedron centred on a 5+ cation bears a net negative charge, which is balanced by a net positive charge on octahedra around the block corners. The whole structure can, in

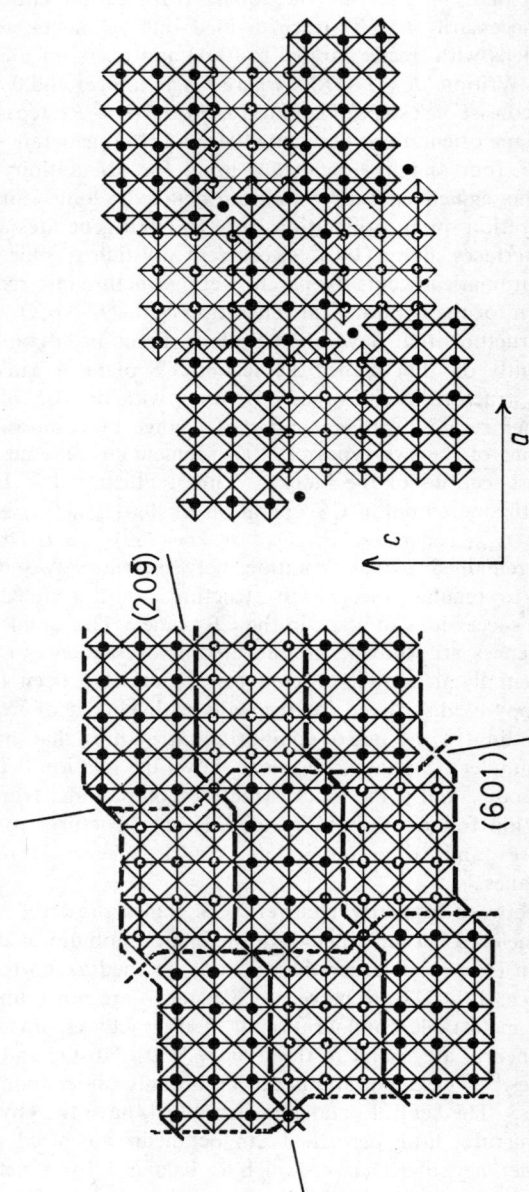

Figure 21 *Generation of the H-Nb_2O_5 structure by collapse of the basic ReO_3 structure along two CS planes, (106) and (209). Anion sites on the CS planes are eliminated; cation sites lying on the intersections of the CS operations move into tetrahedral co-ordination*

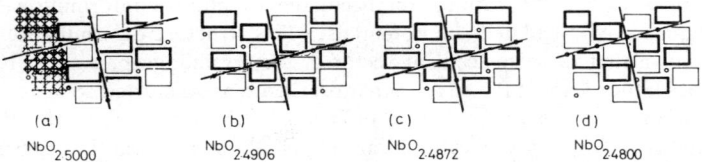

(a) NbO$_{2.5000}$ (b) NbO$_{2.4906}$ (c) NbO$_{2.4872}$ (d) NbO$_{2.4800}$

Figure 22 *Change of composition and generation of regular intergrowths as CS planes swing round in block structures*: (a) H-Nb$_2$O$_5$, CS on (20$\bar{9}$) of parent ReO$_3$ structure (CS plane at 12.53° to (001)$_{ReO_3}$); (b) 1 : 1 intergrowth Nb$_{53}$O$_{132}$, CS on (40$\bar{17}$) (13.24°); (c) 1 : 2 intergrowth Nb$_{39}$O$_{97}$, CS on (50$\bar{21}$)(13.29°); (d) Nb$_{25}$O$_{62}$, CS on (10$\bar{4}$)(14.04°). The CS plane on (601) is common to all

Table 2 *Generation of block structure types from ReO$_3$ structure by double crystallographic shear*

Compound	Block structure	Generating crystallographic shear	
		$(\frac{1}{2}a)[\bar{1}10]/$	$(\frac{1}{2}a)[011]/$
Ti$_2$Nb$_2$O$_7$	$(3 \times 3)_\infty$	(100)	(10$\bar{3}$)
Nb$_{12}$O$_{29}$	$(3 \times 4)_\infty$	(100)	(10$\bar{4}$)
Nb$_{16}$O$_{40}$	$(4 \times 4)_\infty$	(100)	(10$\bar{4}$)
MgNb$_{14}$O$_{35}$F$_2$	$(3 \times 5)_\infty$	(100)	(10$\bar{5}$)
Nb$_{22}$O$_{54}$	$\{(3 \times 3)_1 + (3 \times 4)_\infty\}$	(601)	(20$\bar{7}$)
Nb$_{47}$O$_{116}$	$\{(3 \times 3)_1 + (3 \times 4)_\infty\} + (3 \times 4)_2$	(601)	(40$\bar{15}$)
Nb$_{25}$O$_{62}$	$(3 \times 4)_2$	(601)	(10$\bar{4}$)
Nb$_{53}$O$_{132}$	$\{(3 \times 4)_1 + (3 \times 5)_\infty\} + (3 \times 4)_2$	(601)	(40$\bar{17}$)
Nb$_{28}$O$_{70}$	$\{(3 \times 4)_1 + (3 \times 5)_\infty\}$	(601)	(20$\bar{9}$)
WNb$_{12}$O$_{33}$	$(3 \times 4)_1$	(301)	(10$\bar{4}$)
W$_3$Nb$_{14}$O$_{44}$	$(4 \times 4)_1$	(401)	(10$\bar{4}$)
W$_5$Nb$_{16}$O$_{55}$	$(4 \times 5)_1$	(401)	(10$\bar{5}$)
W$_8$Nb$_{18}$O$_{69}$	$(5 \times 5)_1$	(501)	(10$\bar{5}$)

fact, be viewed as a packing of extended rod-like anionic and cationic elements, and would be destabilized by excessive charge separation. Thus the hypothetical $(6 \times 3)_\infty$ structure for Nb$_2$O$_5$, Nb$_{18}$O$_{45}$, appears to be outside the limits of stability, and has not been obtained. Detailed calculations of site potentials in block structures[43] modify the picture to some extent, but confirm that there are marked differences in site potential for cations within blocks and at corner and edge sites, which should lead to strong site-preference effects when ions of different charge are introduced. Larger blocks, with bigger elements of pure ReO$_3$ structure, can be formed only (*a*) if central cation sites can be occupied by ions of higher charge state, *e.g.* W^{6+}, or (*b*) if anion sites in central octahedra can be occupied by F$^-$ or OH$^-$ in place of O^{2-}. This accords with the observations. A redetermination of the crystal structures of Ti$_2$Nb$_{10}$O$_{29}$ and TiNb$_2$O$_7$ by neutron diffraction[44] has confirmed that there is indeed a strongly preferential occupation of corner

[43] A. K. Cheetham and R. Von Dreele, *Nature Phys. Sci.*, 1973, **244**, 139.
[44] A. K. Cheetham and R. Von Dreele, *Proc. Roy. Soc.*, in the press.

and edge sites by Ti^{4+} cations, the occupancy fractions conforming well to the sequence of calculated site potentials. This structure determination has also shown that the displacement vector of the crystallographic shear is not an exact lattice vector; the edge-shared octahedra are distorted so as to open out considerably the cation–cation distances at the interface, to the extent that the two close rows of cations can be just about resolved, in projection, in lattice images (see Figure 23).

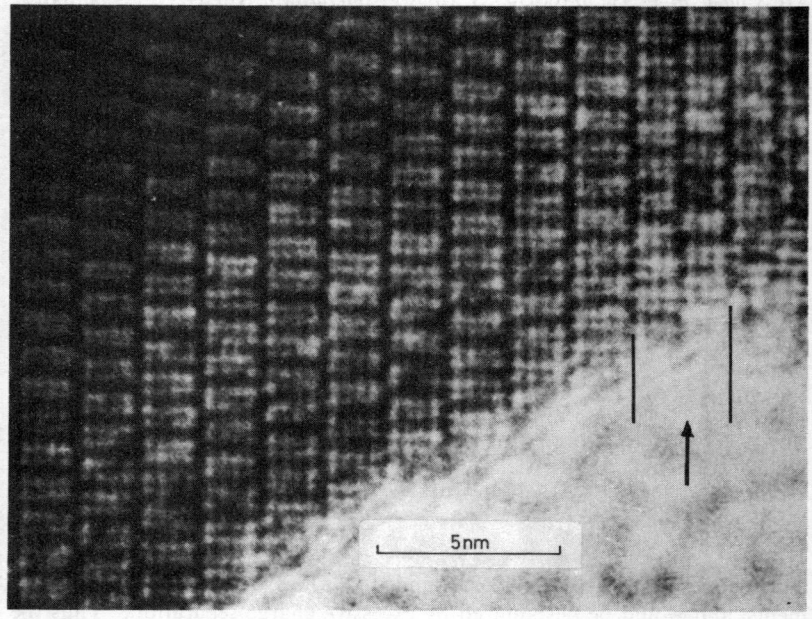

Figure 23 *Lattice image of the* $MgNb_{14}O_{35}F_2$ *phase in the system* MgF_2–Nb_2O_5 *with a Wadsley defect (arrowed) consisting of two rows of intergrown* $Mg_2Nb_{10}O_{25}F_4$ *structure*

Non-stoicheiometry and Defects in Block Structures.—As was made clear in the previous Report, lattice fringe imaging gave clear evidence that deviations from ideal stoicheiometry, and apparently bivariant thermodynamic behaviour of crystalline preparations, must be regarded as arising largely, if not entirely, from coherent intergrowth between two structures that were topologically compatible, but of different composition, *e.g.* $Nb_{25}O_{62}$ and H-Nb_2O_5, in the pseudo-bivariant range $NbO_{2.480}$—$NbO_{2.500}$.[33] Such intergrowths might be isolated lamellae one block wide (Wadsley defects) or more extended domains. The inferences drawn from relatively low-resolution electron microscopy have been fully confirmed by developments in high-resolution microscopy, which enable the $[NbO_6]$ octahedra to be counted individually. Figure 23 shows in this way a defect with two rows of $M_{12}X_{29}$ structure intergrown in the $M_{15}X_{37}$ structure of $MgNb_{14}O_{35}F_2$.

The nature of thermal disorder in block structures has, as yet, received little attention, but Iijima[23] has recorded some suggestive observations on H-Nb_2O_5 quenched from the melt. For pure material, heated in air, the composition is necessarily $NbO_{2.5000}$. Nevertheless, in addition to domains of an alternative, metastable $(3 \times 3)_1$ block structure (crystallographic formula $Nb_{10}O_{25}$) which has also been obtained and characterized in a different way,[34] Iijima unambiguously identified lamellae of $(4 \times 3)_1$ blocks, i.e. of composition $Nb_{13}O_{33}$, $NbO_{2.538}$, which were in parts regularly spaced to form a long-period superstructure. Such oxygen-excess structure elements must be compensated by elements of some hypostoicheiometric structure, in equivalent amount. Compensating structural rearrangements, e.g. elements of $Nb_{25}O_{62}$ structure, were not always seen in association with these defects and the probable way in which a localized metal-excess can be incorporated is considered below. These observations do not relate to the statistical thermodynamic point-defect disorder, except to suggest that this is small, but are important as showing that local disorder transformations within a stoicheiometric crystal may involve adjustments of atomic positions, just as do the extrinsic faults already recognized, conforming always to the crystal chemical constraints of the block structures.

It is known that, in the complex polymorphism of Nb_2O_5,[45] crystallization in some of the metastable forms [e.g. the $(4 \times 4)_\infty$ structure of N-Nb_2O_5] is caused or promoted in the presence of certain dopants—fluorides, water (as source of OH^- anions), SnO_2, etc.—which are incorporated in very small amounts in the products. It is usually considered that dopant ions, of different charge from the host species that they replace, will be compensated by point defects: anion vacancies or interstitial cations would be needed to compensate the replacement of Nb^{5+} by Sn^{4+}. However, the doped niobium oxides have been found to grow with a domain structure, with regions of perfect structure separated by walls consisting of blocks of different size, and hence of different composition.[46] The anion : cation ratios in these domain walls, deduced from their block dimensions, indicate that they correspond to phases incorporating the dopant atoms. It would therefore appear that, at least in growth by vapour transport mechanisms, only restricted amounts of altervalent ions are incorporated, with compensating defects, in the structure of the product crystal. As growth proceeds from an adsorbed layer, the 'impurity' atoms are, from time to time, fully assimilated into elements of a new phase, with its own structure and composition, that can intergrow in perfect coherence with the main product. These observations have a bearing on more general problems of the mechanism of crystal growth—as, indeed, have many of the results emerging from lattice-imaging microscopy—and the facts could

---

[45] H. Schäfer, R. Gruehn, and F. Schulte, *Angew. Chem. Internat. Edn.*, 1966, **5**, 40; A. Reisman and F. Holtzberg, *J. Amer. Chem. Soc.*, 1959, **81**, 3182.

[46] J. S. Anderson, J. M. Browne, and J. L. Hutchison, *Nature*, 1972, **237**, 153.

in this case be detected and interpreted because of the peculiar suitability of block structures for study by lattice-imaging methods, but they may be indicative of the structural response of crystals to dopant atoms and have a wider significance.

Non-stoicheiometric Block Structures.—Reference was made in the last Report to the question whether certain of the block structure compounds have a genuine stoicheiometric range, or whether their apparent variability of composition could be accounted for entirely by the occurrence of Wadsley defects.

How far the apparent stoicheiometric existence ranges of block structures can be fully explained in terms of Wadsley defects is considered below. Systematic studies using chemical techniques and X-ray diffraction, such as those of Kimura[47] on the high-temperature thermodynamics of the NbO_2–Nb_2O_5 system, and of Gruehn[48] on the system Nb_2O_5–Nb_3O_7F, need to be supplemented by electron microscopy, but it is now possible to predict the types of Wadsley defect that may be expected. The approximate composition ranges found for phases in the oxyfluoride system are shown in Table 3. The basic structural components of these phases are the analogues of the oxides between $Nb_{25}O_{62}$ and H-Nb_2O_5, but involving blocks (3×5) and (3×6) in dimensions; one generating CS operation is on $(106)_{ReO_3}$, so that they can form coherent intergrowths with each other and with H-Nb_2O_5; the other CS plane swings round from $(50\bar{1})$ to $(60\bar{1})$. The existence ranges are then consistent with incorporation of Wadsley defects of structure III in H-Nb_2O_5, aperiodically and up to a substantial concentration. Beyond some limit, ordered domains of the regular 1 : 1 intergrowth structure II are formed. Phase III similarly should build in defects of structure I in its substoicheiometric range [although, by analogy with the $(3 \times 4)_2$ structure of $Nb_{25}O_{62}$ a second mode of non-stoicheiometry is possible, as is discussed later]; Wadsley defects of structure V account for the excess-anion range. Similar considerations apply to the preparations with compositions between the limit of III and the ideal composition of V. Above that, between $NbX_{2.529}$ and $NbX_{2.532}$, extra rows of $(5 \times 3)_1$ structure are probably intergrown, even though the corresponding pure oxyfluoride, $Nb_{16}O_{39}F_2$, is not known; the isostructural $MoNb_{15}O_{40}F$ has been characterized. Both intergrowth phases II and IV appear as phases of fixed composition in X-ray diffraction; evidently where lamellae of two compatible structures are present in comparable, but irrational amounts, ordering processes maximize the size of domains of the simplest, regularly alternating sequence.

In a comprehensive summary of block structures, Gruehn[49] pointed out that careful chemical and X-ray-diffraction (Guinier camera) work indicated that those phases which had cations in tetrahedral sites, and only those,

---

[47] S. Kimura, *J. Solid State Chem.*, 1973, **6**, 438.
[48] R. Gruehn, *Z. anorg. Chem.*, 1973, **395**, 181.
[49] R. Gruehn, N.B.S. Special Publication 364, Solid State Chemistry, 1972, p. 63.

Table 3 Apparent stoicheiometric ranges in niobium oxide fluoride structures

	Phase	Structure	Composition, x in NbX$_x$ Ideal	Composition, x in NbX$_x$ Observed	Probable Wadsley defects
I	Nb$_{28}$O$_{70}$	$(3 \times 4)_1 + (3 \times 5)_\infty$	2.5000	$2.500 \leqslant x \leqslant 2.504$	Structure III
II	Nb$_{59}$O$_{147}$F	$\{(3 \times 4)_1 + (3 \times 5)_\infty\} + \{(3 \times 5)_2\}$ 1:1 intergrowth	2.5085	2.510	—
III	Nb$_{31}$O$_{77}$F	$(3 \times 5)_2$	2.5161	$2.515 \leqslant x < 2.520$	{Structure I, Structure V}
IV	Nb$_{65}$O$_{161}$F$_3$	$\{(3 \times 5)_2\} + \{(3 \times 5)_1 + (3 \times 6)_\infty\}$ 1:1 intergrowth	2.5231	2.523	—
V	Nb$_{34}$O$_{84}$F$_2$	$(3 \times 5)_1 + (3 \times 6)_\infty$	2.5294	$2.526 < x < 2.535$	{Structure III, Structure VI}
VI	[Nb$_{16}$O$_{39}$F$_2$]	$(3 \times 5)_1$	2.5625	Hypothetical	

showed well substantiated non-stoicheiometric behaviour; the 'ribbon' structures such as $Ti_2Nb_2O_7$, $(3 \times 3)_\infty$, and $Nb_{12}O_{29}$, $(3 \times 4)_\infty$, were of ideal, invariable composition in monophasic preparations. For example, $TiNb_{24}O_{62}$ and $Nb_{25}O_{62}$ appear to have a range of composition on the metal-rich side (Table 4) which does not include the ideal composition; preparations with the exact composition $M_{25}O_{62}$ ($MO_{2.480}$) are invariably biphasic in X-ray diffraction. Although coherently intergrown single strips or narrow lamellar domains of a second structure would not be detected by X-ray methods, the distinction between compounds with and without tetrahedral sites supports the contention that the observations are genuine. Further evidence, from electron-microscope study of local structure, comes from the work of Allpress and Roth[50] on the Nb_2O_5–WO_3 system. $WNb_{12}O_{33}$ [$(3 \times 4)_1$] and H-Nb_2O_5 [$(3 \times 4)_1 + (3 \times 5)_\infty$] can both intergrow coherently and form a succession of regular intergrowth phases. An excess of cations, above the ideal formula $MO_{2.5385}$, could therefore be accommodated by intergrowing occasional Wadsley defects with the H-Nb_2O_5 structure. A thorough study of the system showed, however, that such Wadsley defects were not present in well annealed preparations between the ideal composition $WO_3,6Nb_2O_5$ and a limit of $WO_3,8Nb_2O_5$. If it is tentatively accepted that the anion sublattice provides the framework for the block structures, this limiting composition could be expressed as $M_{13.05}O_{33}$, with $(3 \times 4)_1$ blocks, as observed, but with a cation excess incorporated in some way.

In addition to this evidence for variability of composition, there are a few compounds for which it is impossible to reconcile the chemical composition with the block structure, as determined, without invoking some form of defect to accommodate a cation excess. One such is a phase to which Levin[51] ascribed the composition $GeO_2,9Nb_2O_5$, which appears to be isostructural with PNb_9O_{25}, a $(3 \times 3)_1$ block structure. A second is a phase with a narrow composition range around $P_2O_5,22Nb_2O_5$ but necessarily with the total stoicheiometry $MO_{2.5000}$;[52] Allpress reported this as having the structure $\{(3 \times 3)_1 + (3 \times 4)_1\}$, corresponding to the ideal formula $M_{23}O_{58}$, again with a cation excess if the real unit-cell composition is expressed as $M_{23.2}O_{58}$. A determination of the structure for such unambiguous cases of metal-excess defect structures would probably furnish a key to the problem of non-stoicheiometric block structures just discussed.

The structure of '$GeO_2,9Nb_2O_5$' has now been determined by both neutron-diffraction and X-ray-diffraction methods.[32] The scattering factors for germanium and niobium differ in opposite senses for neutrons and for electrons; neutron diffraction is particularly suited for determining the occupation fractions of oxygen sites, since the scattering lengths of all the atoms in the crystal are comparable in magnitude. Accepting Levin's composition for the compound, extreme possibilities would be that the

---

[50] J. G. Allpress and R. S. Roth, *J. Solid State Chem.*, 1971, **3**, 209.
[51] E. M. Levin, *J. Res. Nat. Bur. Stand.*, Sect. A, 1966, **70**, 5.
[52] E. M. Levin and R. S. Roth, *J. Solid State Chem.*, 1970, **2**, 250.

Table 4 Composition ranges reported for block-structure oxides with tetrahedral sites occupied

Oxide	O : M Ratio Ideal	O : M Ratio Observed limits	Composition for perfect column structure	Maximum occupancy number for tunnel sites
$WNb_{12}O_{33}$	2.5385	2.529—2.538	$M_{13.05}O_{33}$	1.05
$Nb_{28}O_{70}$	2.5000	2.490 or 2.495 —2.500	$M_{28.12}O_{70}$ or $M_{28.06}O_{70}$	1.12—1.06
$TiNb_{52}O_{132}$ $Nb_{53}O_{132}$	2.4906	2.483	$M_{53.16}O_{132}$	1.08
$TiNb_{24}O_{62}$ $Nb_{25}O_{62}$	2.4800	2.472—2.478	$M_{25.08}O_{62}$— $M_{25.02}O_{62}$	1.08—1.02
$Nb_{47}O_{116}$	2.4681	2.460 2.464—2.467	$M_{47.15}O_{116}$ $M_{47.08}O_{116}$— $M_{47.02}O_{116}$	1.075 1.04—1.01
$Nb_{22}O_{54}$	2.4545	2.453	$M_{22.014}O_{54}$	1.01
'$GeO_2,9Nb_2O_5$'	2.5000	2.474	$M_{10.11}O_{25}$	1.11
'$P_2O_5,22Nb_2O_5$'	2.521	2.5000 (10)	$M_{23.20}O_{58}$	1.06

unit-cell composition would be $M_{10}O_{24.74}$ if the cation sublattice were perfect, or $M_{10.11}O_{25}$ if the oxygen sublattice were perfect. There must be either oxygen 'vacancies' or metal 'interstitials', or both. From the highly concordant structure determinations it emerged (a) that, within narrow error limits, all oxygen sites were fully occupied and (b) that all the octahedral cation sites within the blocks were fully occupied, and occupied by niobium atoms. It follows that the main structure of the crystal, provided by the interfaced columns of $[NbO_6]$ octahedra, is essentially perfect; the formula can be written as $(Nb,Ge)_{1.11}(Nb_9O_{25})$, with the cation excess wholly associated with the way in which possible cation sites are occupied in the channels formed by the corners of the columns (Figure 24). It is reasonable inference that this principle is more widely applicable to the problem discussed in the beginning of this section. The structure determinations showed, in addition, that not all the cations in the channels were on tetrahedral sites but also occupied sites in the corners of the channels which, as close examination of the structure shows, places them in octahedral co-ordination, as outlying appendages to the square columns (Figure 24b). Such potential octahedral sites occur in pairs, at each level $z = 0$, $z = \frac{1}{2}$, and although the insertion of a pair of cations in these sites inevitably blocks the occupation of one tetrahedral site, it provides a means of 'over-filling' the channels. This clearly occurs, to the extent of about 10%. The problem is why it does not proceed further, up to the structurally limited end-composition $M_{11}O_{25}$, corresponding to complete occupation of octahedral sites and elimination of tetrahedral cations. '$GeO_2,9Nb_2O_5$' may have a narrow composition range on the Nb_2O_5-rich side (virtually a solid solution with $Nb_{10}O_{25}$, as is the case for PNb_9O_{25}), but the GeO_2-rich limit is certainly close to the composition assigned by Levin. Several reasons for the restriction on octahedral site occupancy could be advanced: (a) that interactions between cations in the channels impose a one-dimensional ordering, as illustrated in Figure 24g; (b) that the distribution of octahedral and tetrahedral sites is linearly randomized, and determined by the statistical thermodynamics of site preference; (c) that the constraint is thermodynamic, and imposed by coexistence with GeO_2 as a separate phase. If the arrangement is ordered, there is evidently little or no correlation between any one channel and its neighbours, since there is no trace of superlattice ordering. This conclusion would be significant for defect tunnel structures in a wider context.

If this solution of one defect structure can be regarded as typical, and extended to the non-stoicheiometric behaviour considered earlier, it would appear that all the facts fit a consistent pattern, with the channels of tetrahedral sites stuffed to a maximum excess of 5—10% (cf. Table 4, columns 5 and 6). That the $M_{25}O_{62}$ structures cannot attain the ideal composition would indicate that the site preference energies are rather finely balanced. Iijima's observations on melt-quenched H-Nb_2O_5 also fall into place. The type of disorder found—whether thermal intrinsic disorder or metastable disorder, frozen into the sample—showed a local reorganization that provided an excess of $(4 \times 3)_1$ blocks. If the associated corner channels had only their normal complement of cations in tetrahedral sites, there would have been oxygen-excess structure elements, $Nb_{13}O_{33}$; with the total composition of the crystal fixed as $NbO_{2.5000}$, it is likely that the compensating defects were 'stuffed' channels, with octahedral sites occupied.

There is, however, some evidence that the interpretation of these phenomena is not yet complete, from Kimura's investigation[47] of the thermodynamics of the NbO_x ($2 \leqslant x \leqslant 2.500$) system at 1300 and 1400 °C. In each experiment, the equilibration with gas buffer mixtures was approached from both sides: by oxidation of NbO_2 and by reduction of Nb_2O_5. Under

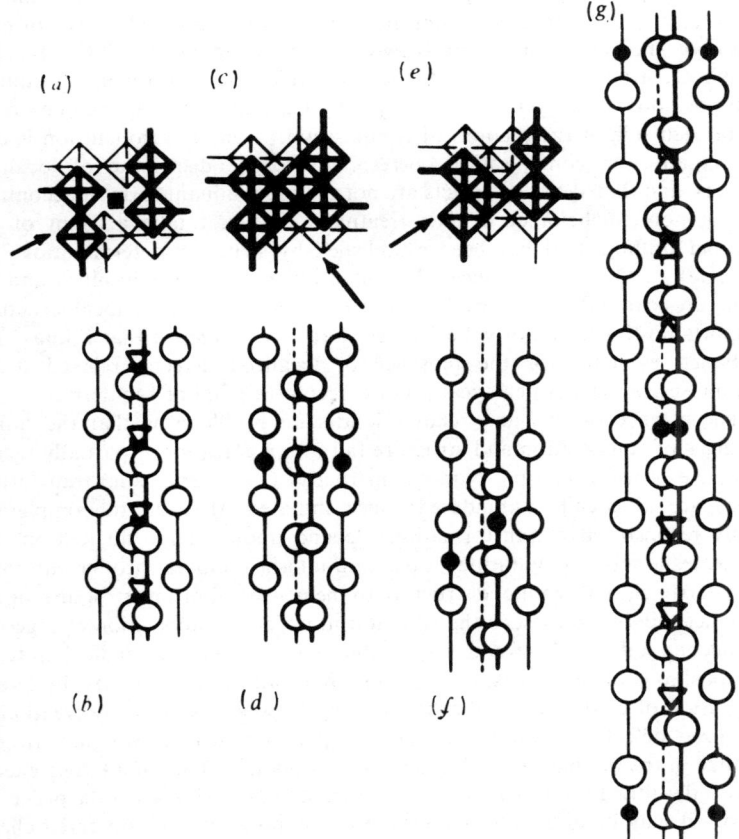

Figure 24 (a), (b): *plan and b-axis projection of normal tetrahedral cation environment at block corners;* (c), (d) *and* (e), (f): *insertions of cations in octahedral co-ordination, replacing tetrahedral sites;* (g): *the possible ordered sequence of tetrahedral and octahedral cations in the corner channels in* $GeO_2,9Nb_2O_5$

these conditions there can be fair confidence that an equilibrium composition and, hopefully, an equilibrium structure, was reached, and on this basis Kimura attempted a thermodynamic analysis in terms of the concentration of four distinguishable types of oxygen atom: those in the block interiors, linked to two six-co-ordinate Nb cations; those in the block edges, co-ordinated to three six-co-ordinate Nb; those at the block corners, with four six-co-ordinate Nb; and those that build up the channels of tetrahedral sites, each linked to one octahedral and one tetrahedral cation. Oxidation and reduction processes were then treated in terms of changes in the concentration of each of these species, the crystal as a whole being regarded as a regular solution system. On this basis, equilibrium constants for the partial reactions governing the concentrations of the different types of oxygen could be derived, and these accounted well not only for the observed ranges of existence of the phases observed between (and including) the $Nb_{22}O_{54}$ and H-Nb_2O_5 structures, but also for the noteworthy fact that the upper limiting compositions of $Nb_{22}O_{54}$, $Nb_{47}O_{116}$, $Nb_{25}O_{62}$, and $Nb_{53}O_{132}$ all lie on the metal-rich side of the theoretical composition. Kimura's conclusion is that the equilibrium configuration corresponds to the dispersion of localized defects, and that Wadsley defects are not the predominant form of accommodating non-stoicheiometry. His treatment does not take account of the 'tunnel stuffing' that has been established by structure determinations for $(Nb,Ge)_{10.11}O_{25}$, but it is clear that the problem of defects in block and CS structures can only be resolved by electron microscopy of the local structure. Reports on this have not yet appeared, but it is of interest that Iijima[53] has obtained evidence for the presence of localized defects—relaxed defect complexes rather than classical point defects—in Kimura's materials.

The essence of this and related solid-state problems is that the linked changes of composition and structure fall into a category of partially reconstructive transformation. Non-reconstructive (*e.g.* martensitic) transformations are kinetically unhindered; once initiated, they go to completion. Fully reconstructive transformations depend upon a large fluctuation that nucleates a new configuration; a reaction interface moves through the solid, transferring atoms from one structure to the new equilibrium structure of the product. Between these is the situation in which a composition change can be accommodated in several ways, none of which requires radical shifts in the positions of the atoms, for example by rotation of CS planes, by lateral migration of CS planes to alter the spacing between them, or by creation of localized defects within the elements of parent structure, with or without change in the orientation and spacing of CS planes. The total lattice energy may differ little for alternative configurations, but the kinetics of the processes involved may be different. Under these circumstances, the observed chemistry and structure of a system may, in reality, be explored over a metastable free energy surface in temperature–composition–chemical potential space.

---

[53] S. Iijima, personal communication; to be published.

5 Infinitely Adaptive Structures

From the phenomenon of pivoting CS planes, discussed in a previous section, and from observations of the superlattice ordering in certain structures, considered below, emerges the concept that, in certain structure types, a fully ordered structure may be formed for every possible composition.[54] Changes in the total atomic ratios of a solid-state system (within a certain range) do not then result in separation into two coexisting phases, neither do they involve the randomized introduction of defects, as in typical non-stoicheiometric solid-solution systems. Instead, the structure adapts its long-range order to define a new crystallographic repeating unit, maintaining a common structural principle.

CS Phases.—It will be clear that the composition of a CS phase can be altered in two ways: by change in the width of the slab of parent structure between CS planes of a given orientation, or by change of orientation of CS planes enclosing slabs of essentially constant width. The former change takes place in discrete steps, generating a homologous series for every CS plane orientation; the latter is, or can be, a continuous process. These two modes are quite independent as regards attainment of order. Variable spacing between CS planes—implying a non-equilibrium departure from fixed stoicheiometry—is frequently found when the lateral spacing (and hence the interaction) between CS planes is large, but observations on rutile- and ReO_3-type CS phases leave no doubt that the equilibrium configuration of any CS surface is planar. This implies that the APB and CS segments in the CS plane take up a regular, repeating sequence.

Considering, for simplicity, CS phases derived from the ReO_3 structure, a compound M_nO_{3n-m} is formed by the omission of every nth $(hk0)$ sheet of anion sites, where m depends upon the CS plane orientation $(hk0)$. The normal spacing between CS planes, d_{CS} is related to n by

$$d_{CS} = d_{hk0}(n - c), \quad (c = \text{fractional collapse})$$

so that the homologous series (n variable) can be represented by a set of composition points on a nearly linear plot of composition against d_{CS}^* ($= 1/d_{CS}$) (Figure 25). A similar array of composition points could be plotted for every orientation; as has been seen, there is a continuum of possibilities. If, now, the orientation be written as $(h'10)$, where $h' = h/k$, the transverse lines with $n = $ constant represent paths by which the crystal could change its composition without introducing or eliminating CS planes. As long as the sequence of segments in the CS planes adjusts itself to a regular repeating pattern, the system remains fully ordered.

[54] J. S. Anderson, *J. C. S. Dalton*, 1973, 1107.

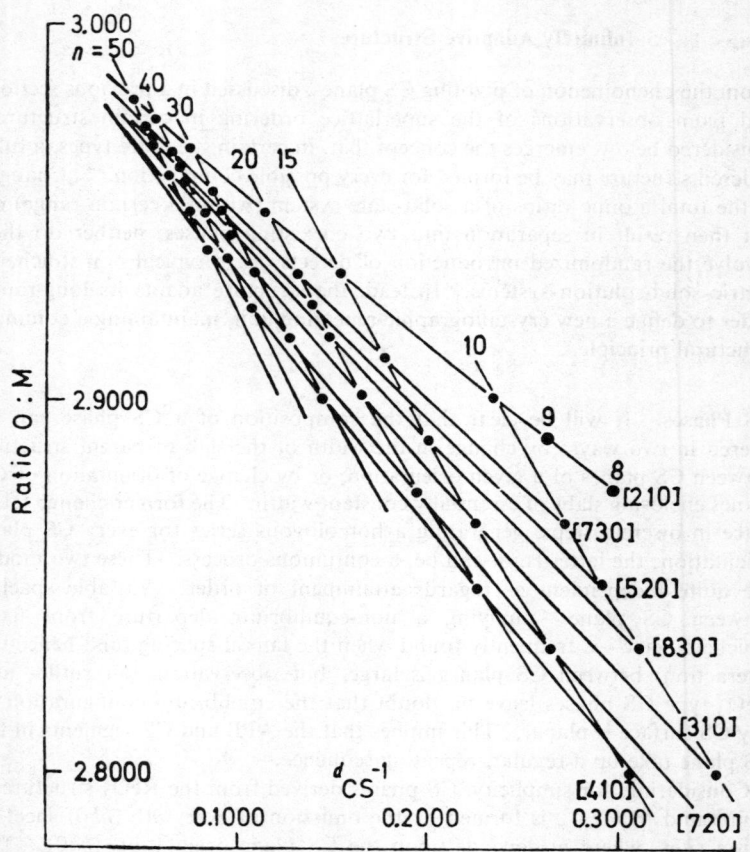

Figure 25 *Change of composition and reciprocal of the spacing between CS planes in ReO$_3$-based structures as the CS plane swings round from* (410) *to* (210)

Adaptive Superlattice Ordering.—A somewhat different, but related, situation is found in certain structures in which a one-dimensional ordering process generates a superlattice from a simpler sub-cell. The occurrence of adaptive ordering in such a system—the low-temperature polymorph of Ta$_2$O$_5$—was established experimentally by Roth and Stephenson[55] before the full implications were recognized. L-Ta$_2$O$_5$ and ternary oxides derived from it have structures related to that of U$_3$O$_8$, basically consisting of ribbons of edge-sharing pentagonal-bipyramidal [MO$_7$] co-ordination groups extending along one crystallographic direction (the b axis). The metal atoms form a pseudo-hexagonal (orthorhombic) array, but the oxygen atom positions and small

---

[55] R. S. Roth and N. C. Stephenson, 'The Chemistry of Extended Defects in Non-metallic Solids', ed. L. Eyring and M. O'Keefe, North-Holland, Amsterdam, 1970, p. 167.

displacements of the metal atoms impose a superstructure of this pseudo-hexagonal subcell, along the b direction. WO_3, ZrO_2, Al_2O_3, and other oxides simulate the formation of solid solutions with L-Ta_2O_5,[56] but careful X-ray-diffraction study reveals that the superstructure lines display a remarkable behaviour. They shift systematically and continuously with changes of composition. Each one of a wide range of compositions between Ta_2O_5 and $4WO_3,11Ta_2O_5$, examined by single-crystal X-ray methods, had a unique superstructure; in no well annealed specimen were two phases, each with its own superstructure, found to coexist in equilibrium.[57] It was necessary to assign high values to the superstructure multiplicity.

The interpretation of these facts, based on single-crystal structure determinations[58] and the related structure of U_3O_8,[59] is that the superstructure arises from the folding or undulation of the ribbons of pentagonal bipyramids, with a periodicity commensurable with a repeating unit of 5, 8, 11 ... $3n+2$ subcell spacings along the b axis (Figure 26). This generates a set of homologous subunits $M_{10}O_{26}$, $M_{16}O_{42}$, $M_{22}O_{58}$ ... $M_{2n}O_{(16n-2)/3}$. These subunits may themselves be stacked along the b axis, in varying proportions but in a repeating sequence, to form a longer superlattice. Denoting the multiplicities of the subunits by m_1, m_2 ... m_i, and the number of each in the true superlattice repeat by a_1, a_2 ... a_i, the total multiplicity m^* may be written as

$$m^* = a_1 m_1 + a_2 m_2 + \cdots = \sum a_i m_i$$

If the compositions of the subunits are denoted by x_1, x_2 ... x_i, the composition of the true repeating unit is $x^* = \sum a_i x_i$. Changes of composition in the phases formed between a pair of oxides (*e.g.* WO_3 and Ta_2O_5) imply changes in chemical potential, and it would appear that these impose the thermodynamic constraint (analogous to the biphasic equilibrium between compounds in an ordinary pseudobinary system) that only adjacent pairs of subunits can exist within any one structure. By changes in the relative proportions of subunits, every possible composition between $M_{10}O_{26}$ and $M_{16}O_{42}$, $M_{16}O_{42}$ and $M_{22}O_{58}$, *etc.* could form, in principle, an ordered structure, with the multiplicities $m^* = 5a_1 + 8a_2$, or $8a_2 + 11a_3$, *etc.* Figure 27 shows the hypothetically continuous variation of composition as the ratio $a_1 : a_2$, $a_2 : a_3$ changes, together with the compositions actually examined and the superstructure multiplicities assigned by Roth and Stephenson.

This is the structural principle. The actual situation in the L-Ta_2O_5 oxide series is rather more complex in that the oxygen : metal ratios, running from 2.49 in Al_2O_3– and ZrO_2–Ta_2O_5 phases to 2.5769 in $WTa_{22}O_{67}$, are all oxygen deficient as compared with the ideal structures (ratio = 2.6000, 2.6250, and 2.6364 for $m = 5$, 8, and 11 respectively). In the real structures,

[56] R. S. Roth and J. L. Waring, *J. Res. Nat. Bur. Stand., Sect. A*, 1970, **74**, 485.
[57] R. S. Roth, J. L. Waring, and H. S. Parker, *J. Solid State Chem.*, 1970, **2**, 445.
[58] N. C. Stephenson and R. S. Roth, *Acta Cryst.*, 1971, **B27**, 1010, 1018, 1025, 1031, 1037.
[59] A. S. Andreasen, *Acta Cryst.*, 1958, **11**, 612; B. O. Loopstra, *ibid.*, 1964, **17**, 651.

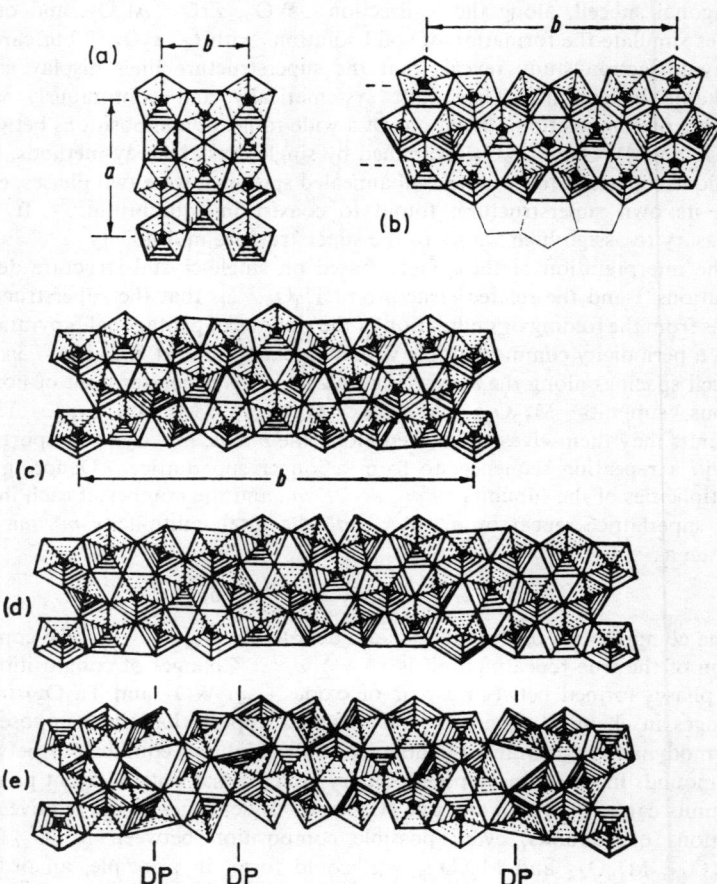

Figure 26 *Structural principle of L-Ta_2O_5 and related compounds*: (a) U_3O_8; (b) $m = 5$ *unit*, $M_{10}O_{16}$, *showing folding of the chain of pentagonal bipyramids*; (c) $m = 8$ *unit*, $M_{16}O_{42}$; (d) $m = 11$ *unit*, $M_{22}O_{58}$; (e) Ta_2O_5, $Ta_{22}O_{55}$ *with three distortion planes (DP) per unit cell*

as illustrated in Figure 26e for the $m^* = 11$ form of Ta_2O_5 itself, some oxygen sites are periodically eliminated, leaving the metal atoms in highly distorted co-ordination polyhedra. Such distortion planes could be introduced, singly or in pairs, at every folding point of the undulating chains of pentagonal bipyramids. A maximum is thereby set to the number of oxygen atoms that can be omitted from each basic subunit, and from each superstructure multiplicity m^* there could, in principle, be derived a homologous series of compounds, with discrete compositions arising from the operation of a

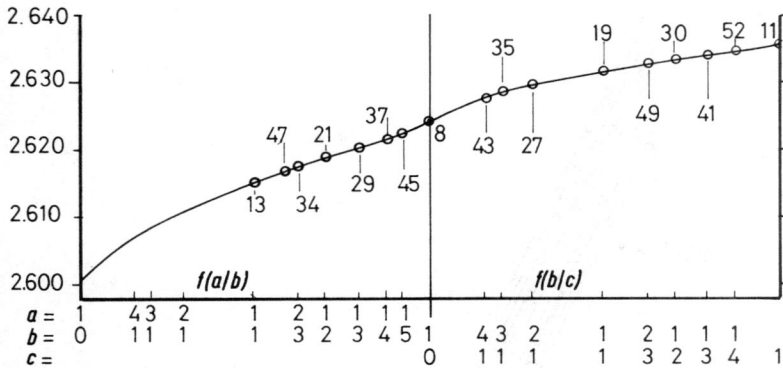

Figure 27 Change of composition in (filled) L-Ta₂O₅-like structures as the ratio of constituent subunits changes. Marked points show total superstructure multiplicities identified in Ta₂O₅–WO₃ series of phases

variable number p of distortion planes per true unit cell, where $0 \leqslant p \leqslant p_{max}$. The general formula for such compounds could be expressed as

$$M_{2m^*}O_{\{2(8m^* - a_1 - a_2)/3\} - p}$$

If the linear density of distortion planes (i.e. d^*_{DP}) is taken as a configurational parameter and plotted against the composition, every multiplicity m^* generates a linear array of composition points. Figure 28 shows this for some of the m^* values identified by Roth and Stephenson. If long-range ordering is indeed capable of arranging the subunits into indefinitely large superstructures, the arrays of composition points form a continuum, and a second mechanism is provided whereby a solid solution or mixed-valence phase could vary continuously in composition without any disorder.

In any particular system (e.g. the Ta₂O₅–WO₃ system as shown in Figure 28, and similarly for phases reported in the Al₂O₃–Ta₂O₅ system,[58] and for a particular temperature of internal equilibration, there is clearly some superstructure sequence and density of distortion planes that represents the most stable configuration, but each composition does not necessarily correspond to a unique structure. In particular, the composition $MO_{2.5000}$ is represented by a composition point for every possible value of m^*—an infinity of potential superlattice structures. There is indeed direct evidence from the crystallographic literature,[60] and from newer electron-microscopic work, that the superlattice of L-Ta₂O₅ changes with the temperature of annealing; the value $m^* = 11$ and structure found by Roth and Stephenson relate specifically to the oxide as equilibrated at ca. 1200 °C.

A related situation is found in the 'solid solutions' formed in the YOF–

---

[60] E.g. R. Moser, *Schweitz. mineral petrograph. Mitt.*, 1965, **45**, 35.

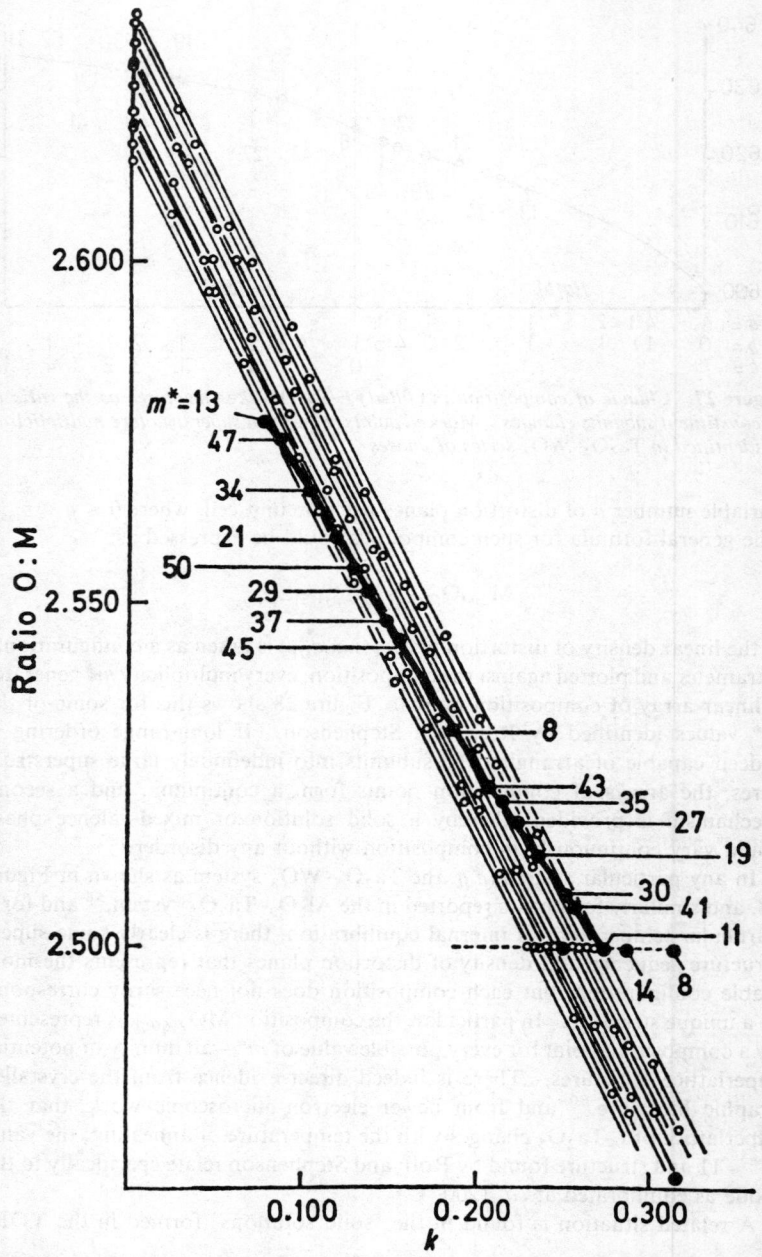

Figure 28 L-Ta_2O_5 superstructures: variation in composition with reciprocal of average position between distortion planes for several observed multiplicities m^*. Points represent compositions, for each m^* value, as successive distortion planes are introduced. The heavy line joins composition points identified in the Ta_2O_5–WO_3 system

YF_3 [61] and the zirconia-rich $ZrO_2-Nb_2O_5$ [62] systems, which provide a one-dimensionally ordered means of accommodating an anion excess in structures related to the fluorite type. Bevan and Mann have interpreted the former in terms of regularly recurrent layers in which all the excess of anions is concentrated, and having a structure related to that of YF_3, interpolated along one axial direction of the (distorted) fluorite structure of YOF. If such a 'stuffed' layer recurs every n fluorite cells, the composition of the ordered structures is $Y_nO_{n-1}F_{n+2}$. Over the composition range $YX_{2.130}$—$YX_{2.220}$ (X = O + F), however, careful X-ray-diffraction work shows the characteristic behaviour already mentioned: the superstructure lines shift continuously with composition; no two compositions, however closely spaced and however irrational, had the same superstructure, neither did any show any splitting of superstructure lines, indicative of two ordered phases coexisting; the observed superstructure lines could, for most preparations, be interpreted only in terms of high multiplicities. These high multiplicities could be interpreted as arising from the regularly recurrent sequence of a set of subunits with individual multiplicities $n = 4, 5, \ldots 8$. The total superstructure multiplicity m^* is then given by $m^* = a_1n_1 + a_2n_2$ and the corresponding composition, $Y_{m^*}X_{2m^*+a_1+a_2}$, can be symbolized $(n_1)_{a_1}(n_2)_{a_2}$. Data for the closely spaced compositions examined by Bevan and Mann are shown in Table 5.

Table 5 Closely spaced superstructure phases in $Y(O,F)_n$

Composition n (experimental*)	Assigned multiplicity	Constitution	C-axis/ Å	Composition n (calculated for structure)
2.13(6)	23	$(8)_2(7)_1$	126.7	2.130
2.11(5)	7	$(7)_1$	38.6	2.143
2.14(8)	19	$(7)_1(6)_2$	104.9	2.153
2.14(9)	45	$(7)_3(6)_4$	248.2	2.155
2.16(5)	6	$(6)_1$	33.1	2.167
2.17(1)	17	$(6)_2(5)_1$	94.0	2.176
2.18(3)	47	$(6)_2(5)_7$	260.0	2.191
2.18(7)	57	$(6)_2(5)_9$	315.4	2.193
2.22(0)	28	$(5)_4(4)_2$	155.2	2.214
2.25(9)	41	$(5)_5(4)_4$	227.4	2.220

* Fourth, bracketed digit uncertain in significance.

For these structures, the composition uniquely determines the mode of ordering. There is no mechanism whereby a given super-superstructure can generate a family of homologous compounds. In this respect the situation is clearer, and the interpretation more direct, than for the L-Ta_2O_5 phases.

The experimental observations hardly permit any other interpretation that that certain phases have what may be termed infinitely adaptive structures.

---

[61] D. J. M. Bevan, ref. 49, p. 479; A. W. Mann and D. J. M. Bevan, *J. Solid State Chem.*, 1972, **5**, 410.
[62] J. L. Galy and R. S. Roth, *J. Solid State Chem.*, 1973, **7**, 277; R. S. Roth, J. L. Waring, W. S. Brouwer, and H. S. Parker, ref. 49, p. 183.

The question of how far a crystal can really adjust its structure to its composition, to exhibit a continuum of ordered structures, has to be considered at two levels. At the philosophical level, the problem is that of attaining long-range order with large repeating units of structure. In the cases cited, the repeating units are up to a few hundred ångströms in length and the nature of the interactions that impose such large repeat patterns is not well understood. It is evident that, as the superlattice dimensions increase, the energetic difference between a perfectly ordered structure and one with mistakes in stacking, or in the sequence of APB and CS steps in a crystallographic shear plane, must become smaller than the activation energy for the requisite rearrangement process. A genuine continuum of perfect structures must be unattainable. Operationally, it is not possible to characterize the ordered structures with sufficient precision to discriminate between very closely spaced ratios of the subunits that determine the stacking multiplicity or the orientation of CS planes. It will be noted that, for L-Ta_2O_5, Roth and Stephenson limited their superstructure assignments to values around or below 50, corresponding to small values for the proportions of the constituent subunits. Similarly, for the Cr_2O_3–TiO_2 CS phases, Bursill, Hyde, and Philp rationalized the orientations of the CS planes to fairly small values for h, k, and l. The precision attainable in measuring the positions of X-ray- and electron-diffraction spots does not permit any other course. In Table 5, the discrepancies between the composition of the material and that calculated from the structure illustrate the problem; for the particular crystals examined, there is little doubt that the calculated compositions are the more reliable.

Although, therefore, real systems must approximate, rather than conform exactly, to the concept of infinitely adaptive structures, the consequences need to be worked out in their bearing on the nature and thermodynamics of solid solution systems, broadly interpreted. The role of ordering processes has become apparent in relation to a rather exceptional group of materials, which exhibit the requisite internal mobility and unspecified long-range interactions. There are other solid systems to which the same considerations may well apply, which are known (for example) to exhibit variable and irrational superlattice ordering, suggestive of a large true repeating unit. The variable-composition $Fe_{1-x}S$ phase now appears to be of this type.[63] Problems of this kind merit careful study; mobile, adaptive superstructure ordering could well have a wider bearing on highly defective solid-solution phases and non-stoicheiometric compounds.

[63] N. Morimoto, H. Nakazawa, K. Nishiguchi, and M. Tokonami, *Science*, 1970, **168**, 964; H. Nakazawa and N. Morimoto, *Materials Res. Bull.*, 1971, **6**, 345.

2
The Geometry of Disclinations in Crystals

BY W. F. HARRIS

1 Introduction

Let it be said at the outset that the disclination is not likely to prove very important in those materials in which the dislocation has provided so much interest, that is in metal crystals. As can be seen in Figures 1 and 2, twist and wedge disclinations, respectively, the lattice planes become strongly curved

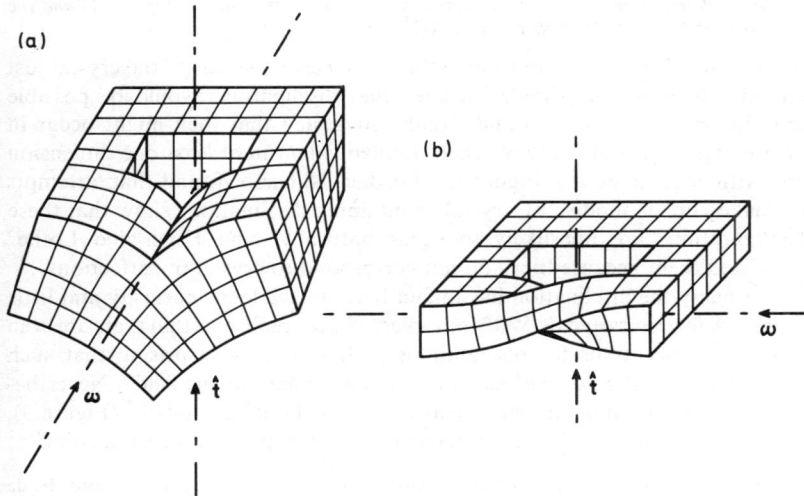

Figure 1 *Twist disclinations of rotation $\pi/2$ rad in a cubic crystal. The rotation ω is perpendicular to the disclination line with unit tangent $\hat{t}$. The disclinations have large hollow cores. In (a) the disclination line intersects ω while in (b) it does not. The twist disclination of (a) could be introduced by cutting into the body in the direction ω as far as $\hat{t}$ and inserting a wedge of angle $\pi/2$ rad from below the figure so that its apex lies along ω. The disclination of (b) could be introduced by making the same cut but simply rotating or twisting the right-hand face of the cut relative to the other through $\pi/2$ rad about ω without inserting or removing any material. Rotating the crystal of (b) through $\pi/2$ rad about $\hat{t}$ shows that the two twist disclinations differ only by the location of ω and not the direction. (a) and (b) correspond to Volterra's distorsions of orders 4 and 5, respectively (cf. Figure 7d and e). (b) is essentially the first of Frank's 'Möbius crystals' (Figure 1 of ref. 3)*

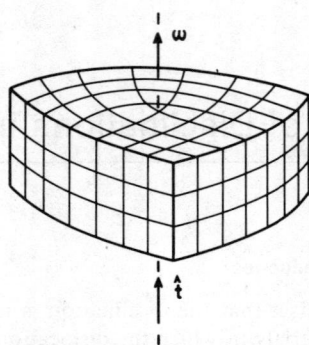

Figure 2 *A positive wedge disclination of rotation $\pi/2$ rad. The rotation ω is parallel to and coincident with the disclination line (unit tangent $\hat{t}$). The defect may be viewed as a rising from the removal of a sector of angle $\pi/2$ rad. The central square on the top surface becomes a triangle* (cf. *Figure* 13). *Insertion of a sector results in a negative wedge disclination* (cf. *Figures* 14 *and* 22) *which has ω and $\hat{t}$ antiparallel. Such a defect corresponds to Volterra's distorsion of order* 6 (cf. *Figure* 7f) *and the second of Frank's* '*Möbius crystals*' (*Figure* 1 *of ref.* 3)

or twisted. The strains are so high that, in Kröner's words,[1] 'the crystal just (could) not bear it'. Friedel[2] argued that disclinations would be possible only in very small crystals and Frank[3] suggested that they might occur in 'pathological crystals' or crystals of limited extent in at least one dimension and with large holes (*e.g.* Figure 1). Further, Nabarro[4] found that 'attempts to make rubber models of crystals containing disclinations show that these configurations are not likely to occur naturally' and Hirth and Lothe[5] dismissed them because they 'do not correspond to crystal imperfections'.

For many the disclination has in fact been an academic curiosity and little more. A few years ago de Wit[6] remarked, 'I now realize... that there are even practical applications for disclinations'. Indeed it is surprising that such unlikely defects should be of any practical importance in any field. Nevertheless they have been described in many systems: liquid crystals[7,8] (Figure 3), the Abrikosov lattice of lines of magnetic flux in type II superconductors[9–11]

---

[1] E. Kröner, in 'Fundamental Aspects of Dislocation Theory', ed. J. A. Simmons, R. de Wit, and R. Bullough, Nat. Bur. Stand. (U.S.A.), Special Publ. 317, 1970, **1**, 712.
[2] J. Friedel, 'Les Dislocations', Gauthier-Villars, Paris, 1956, p. 4.
[3] F. C. Frank, *Phil. Mag.*, 1951, **42**, 809.
[4] F. R. N. Nabarro, 'Theory of Crystal Dislocations', Clarendon Press, Oxford, 1967, p. 125.
[5] J. P. Hirth and J. Lothe, 'Theory of Dislocations', McGraw-Hill, New York, 1968, p. 4.
[6] R. de Wit, ref. 1, p. 710.
[7] J. Friedel and M. Kléman, ref. 1, p. 607.
[8] F. R. N. Nabarro, *J. Phys.*, Paris, 1972, **33**, 1089.
[9] K. H. Anthony, U. Essmann, A. Seeger, and H. Träuble, in 'Mechanics of Generalized Continua' (IUTAM Symposium, Freundenstadt-Stuttgart, 1967), ed. E. Kröner, Springer-Verlag, Berlin, 1968, p. 355.
[10] H. Träuble and U. Essmann, *Phys. Status Solidi*, 1968, **25**, 373.
[11] K. H. Anthony, ref. 1, p. 713; *Arch. Rat. Mech. Anal.*, 1970, **39**, 43.

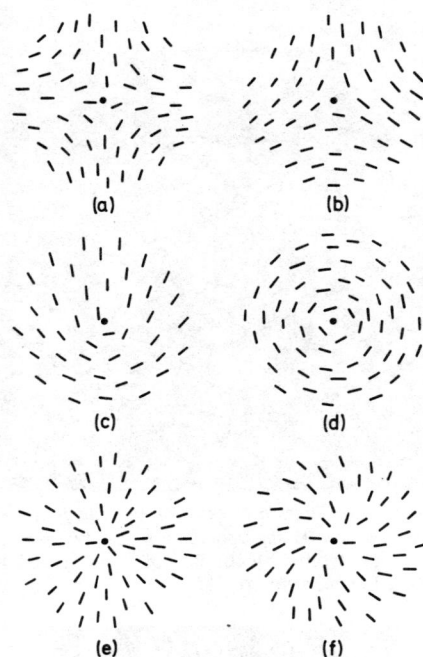

Figure 3 *Wedge disclinations in nematic liquid crystals, after Frank, ref.* 34. (a) *and* (b) *are negative wedge disclinations of rotation* 2π *and* π rad, *respectively;* (c) *is a positive wedge disclination of rotation* π rad; (d)—(f) *are different configurations around positive wedge disclinations of rotation* 2π rad. *A twist disclination of rotation* π rad *is shown in Figure* 30b

(Figure 4), the vector field of spins in magnetic systems,[8,12] the lattice of Bloch walls in ferromagnets[13] (Figure 5), geological strata,[14] polymers,[15,16] and biological structures of many kinds.[17–19] It has been suggested that they may be found in arrays of dislocations[20] and that they may be useful in the study of isotropic fluids and glasses.[21] Disclinations have even been

[12] M. Kléman, *Phil. Mag.*, 1970, **22**, 739.
[13] J. P. Hirth and R. G. Wells, *J. Appl. Phys.*, 1970, **41**, 5250.
[14] J. M. Galligan, *Nature*, 1972, **240**, 144.
[15] J. C. M. Li and J. J. Gilman, *J. Appl. Phys.*, 1970, **41**, 4248; J. J. Gilman, *J. Appl. Phys.*, 1973, **44**, 2233.
[16] K. H. Anthony and E. Kröner, *Koll. Z. Z. Polym.*, in press; in 'Batelle Conference on Polymers', Kronberg, Germany, in press.
[17] W. F. Harris, Thesis, University of Minnesota, Minneapolis, 1970.
[18] Y. Bouligand, *J. Phys.*, Paris, 1969, **30**, C4-90.
[19] Y. Bouligand, *Tissue Cell*, 1972, **4**, 189.
[20] J. D. Eshelby, ref. 1, p. 715.
[21] S. I. Ben-Abraham, ref. 1, p. 717.

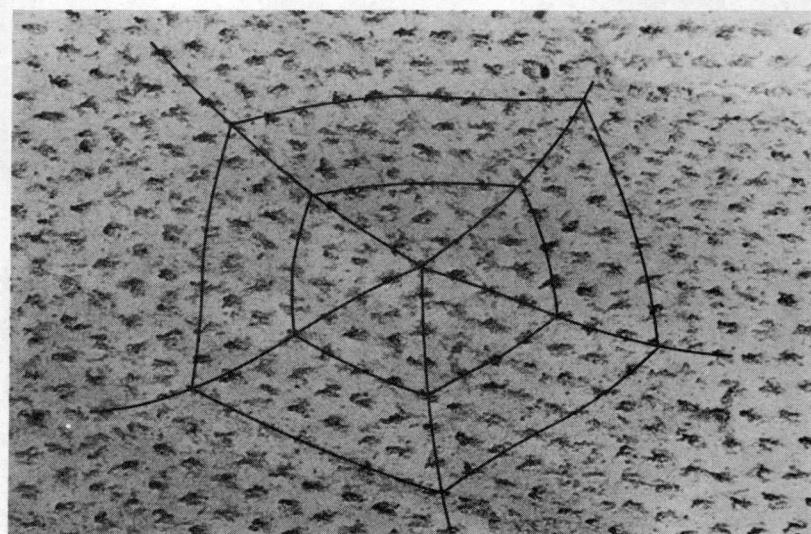

Figure 4 *A positive wedge disclination of rotation* $\pi/3$ rad *in an Abrikosov lattice of lines of magnetic flux in a type* II *superconductor* (cf. *Figures* 18 *and* 19) (Reproduced by permission from 'Mechanics of Generalized Continua', ed. E. Kröner, Springer-Verlag, Berlin, 1968, p. 355)

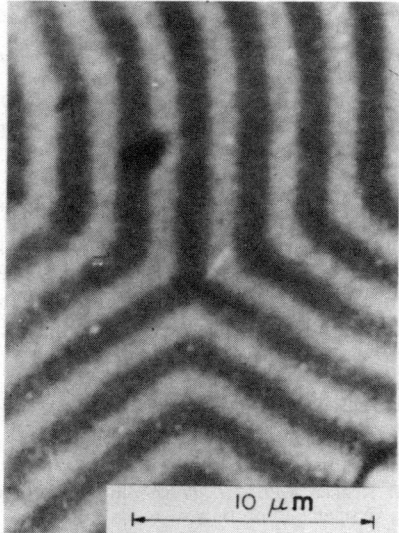

Figure 5 *A negative wedge disclination of rotation* π rad *in the lattice of Bloch walls in a ferromagnet* ($BaFe_{12}O_{19}$). *The basal plane* (0001) *is in the plane of the paper* (Reproduced by permission from *J. Appl. Phys.*, 1970, **41**, 5250)

used by architects.[22,23] To the writer's knowledge no disclination has been found in the lattice of a conventional crystal of non-biological origin.

The above remarks apply to *total* or *perfect* disclinations. *Partial*[24] disclinations, analogues of *partial* dislocations, do in fact occur in conventional crystals including metal crystals in particular. They are associated with terminations, within a crystal, of grain[25] and twin[26] boundaries. Partial wedge disclinations of rotation 0.1280 rad lie along the axes of fivefold rotational symmetry in small twinned fcc crystals known as pseudopentagonal twins.[27-31] These particular partial disclinations de Wit[28] terms appropriately *star disclinations*. Unless otherwise qualified the term *disclination*, in this chapter, will refer to *total* disclinations only. Similarly *dislocation* and *dispiration* are restricted to the *total* varieties of the defects.

Historical Remarks.—The term disclination was first used in print by Nabarro[32] in 1966 though it had been coined earlier by Frank.[33] Frank[34] had originally used *disinclination* but, as Nabarro[35] explains, had dropped it when told by a professor of English that he would be disinclined to use that word. Disinclination survives in its French equivalent, *disinclinaison*.[8,36,37] *Rotation dislocation*[7,38,39] and *rotational dislocation*[40] are other synonyms occasionally found. Wedge disclinations have been referred to as *wedge dislocations*.[41]

The concept, however, is much older than the name. It has at least six independent origins. In crystals translationally periodic in three dimensions it appears first in Frank's 'Möbius crystals'.[3] (Figure 1 of ref. 3 shows

[22] R. W. Marks, 'The Dymaxion World of Buckminster Fuller', Reinhold Publ. Corp., New York, 1960, Figures O23, Q3, T4, T8; many other examples in 'Domebook 2', Shelter Publications, Bolinas, California, 1972.
[23] R. B. Fuller, 'Ideas and Integrities', Collier-Macmillan Canada, Toronto, 1969, opposite p. 193; originally published by Prentice-Hall, New York, 1963.
[24] This terminology appears first in ref. 4, p. 127.
[25] Ref. 4, p. 48.
[26] Ref. 4, p. 127.
[27] R. de Wit, *J. Appl. Phys.*, 1971, **42**, 3304.
[28] R. de Wit, *J. Phys.* (C), 1972, **5**, 529.
[29] J. M. Galligan, *Scripta Met.*, 1972, **6**, 161; *Phys. Letters* (A), 1972, **39**, 407.
[30] T.-W. Chou, *Scripta Met.*, 1973, **7**, 151.
[31] N. Uyeda, M. Nishino, and E. Suito, *J. Coll. Interface Sci.*, 1973, **43**, 264.
[32] F. R. N. Nabarro, in 'International Congress on Electron Diffraction and Crystal Defects' (Melbourne, 1965), Austr. Acad. Sci., Melbourne, and Pergamon Press, Oxford, 1966, p. II L-1.
[33] Ref. 4, p. 20.
[34] F. C. Frank, *Discuss. Faraday Soc.*, 1958, **25**, 19.
[35] F. R. N. Nabarro, ref. 1, p. 710.
[36] M. Kléman and J. Friedel, *J. Phys.*, Paris, 1969, **30**, C4-43.
[37] Y. Bouligand and M. Kléman, *J. Phys.*, Paris, 1970, **31**, 1041.
[38] J. Friedel, 'Dislocations', Pergamon Press, Oxford, 1964, pp. 6, 9, 26.
[39] H. Schaefer, ref. 9, p. 57.
[40] Ref. 38, p. 119.
[41] J. D. Eshelby, W. T. Read, and W. Shockley, *Acta Met.*, 1953, **1**, 251; R. W. Armstrong, *Science*, 1968, **162**, 799.

drawings of twist and negative wedge disclinations of rotation $\pi/2$ rad.[42]) In the meantime Rosin,[43] in his analysis of fibrillar patterns in the epidermis of amphibian larvae, had illustrated positive and negative wedge disclinations of rotation $n\pi/2$ rad for several positive integers n. He referred to them as *Ausnahmepunkte* (see also ref. 44). Rosin's index r is related to the rotation of the disclination by

$$\omega = \pm(2 - r/2)\pi \text{ rad} \tag{1}$$

where the positive and negative signs apply to positive and negative wedge disclinations, respectively. In elastic continua the disclination can be traced to Weingarten[45] and Volterra.[46,47] Volterra's *distorsions*[47] of orders 4 and 5 are twist disclinations and the one of order 6 is a wedge disclination (Figure 7d—f). In liquid crystals disclinations were first reported by Lehmann[48] in 1889 and described in greater detail later.[49–51] Lehmann published colour illustrations of these defects and referred to them as *Symmetriepunkte* or *Kernpunkte* and *Convergenzpunkte*.[49] His work was followed up by Friedel and other French workers.[52–54] Wedge disclination lines seen end on as in Figure 3 Friedel[54] called *noyaux* while disclinations seen from the side he termed *fils*. He assigned an index f ($\omega = |f|\pi$ rad) and found examples of positive ($f = 1,2$) and negative ($f = -1, -2$) wedge disclinations. The same defects were studied later by Oseen,[55] Frank,[34] who termed them *disinclinations*, and Robinson and co-workers.[56] Among the earliest reports of disclinations are the descriptions of singularities (loops and whorls are positive wedge disclinations of rotation π and 2π rad, respectively, while deltas or triradii are negative wedge disclinations of rotation π rad) in fingerprints.[57–59]

[42] Also often called a wedge disclination of rotation $-\pi/2$ rad. This chapter employs the convention that the rotation ω is always positive. The two classes of wedge disclination are distinguished by the adjectives *positive* and *negative*.
[43] S. Rosin, *Rev. Suisse Zool.*, 1946, **53**, 133.
[44] L. Picken, 'The Organization of Cells and Other Organisms', Clarendon Press, Oxford, 1962, Figures 88—90.
[45] G. Weingarten, *Rend. R. Acc. Lincei, Roma*, [5], 1901, **10**, 57.
[46] V. Volterra, *Rend. R. Acc. Lincei, Roma*, [5], 1905, **14**, 193.
[47] V. Volterra, *Ann. Sci. École Norm. Supér., Paris*, [3], 1907, **24**, 401.
[48] O. Lehmann, *Z. phys. Chem.*, 1889, **4**, 462.
[49] O. Lehmann, *Ann. Physik*, 1900, **2**, 649; see also the colour illustrations at the end of the volume.
[50] O. Lehmann, 'Flüssige Kristalle', Engelmann, Leipzig, 1904.
[51] O. Lehmann, 'Die neue Welt der flüssigen Kristalle', Akademische Verlagsgesellschaft m.b.H., Leipzig, 1911.
[52] G. Friedel and F. Grandjean, *Bull. Soc. franç. Min. Crist.*, 1910, **33**, 192 and 409.
[53] C. Mauguin, *Bull. Soc. franç. Min. Crist.*, 1911, **34**, 71.
[54] G. Friedel, *Ann. Physique*, 1922, **18**, 273.
[55] C. W. Oseen, *Trans. Faraday Soc.*, 1933, **29**, 883.
[56] C. Robinson, J. C. Ward, and R. B. Beevers, *Discuss. Faraday Soc.*, 1958, **25**, 29.
[57] H. Faulds, *Nature*, 1880, **22**, 605.
[58] F. Galton, *Nature*, 1888, **38**, 201.
[59] F. Galton, 'Finger Prints', Macmillan, London, 1892.

The Geometry of Disclinations in Crystals

Disclination-like singularities can also be found in the work of mathematicians of the nineteenth century (*e.g.* refs 60—62).

Yet another independent origin of the disclination lies in the architectural work of Buckminster Fuller.[22,23,63] Wedge disclinations are necessary structural components of Fuller's geodesic domes. His work is an important part of the history of disclinations because it provided the inspiration for the determination of the structure of the protein shell or capsid of spherical viruses by Caspar and Klug.[64,65] In these viruses positive wedge disclinations of rotation $\pi/3$ rad are situated at the *pentamers* or *pentons*[66] (*cf.* Figures 19, 21).

2 Weingarten–Volterra Dislocations in Isotropic Bodies

Figure 6 shows a ring of isotropic elastic material with a radial cut C and axis $\hat{t}$. Starting at some point away from C and performing a circuit in a right-handed sense, that is clockwise when looking in the $\hat{t}$ direction, one encounters first one face, C^+, of the cut and then the other C^-. Face C^+ is translated parallel to the axis $\hat{t}$ of the ring relative to face C^- and the two faces are joined.

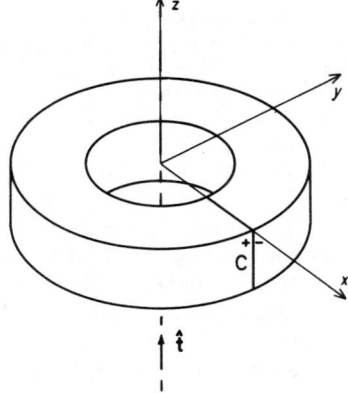

Figure 6 *A doubly-connected body rendered singly-connected by a radial cut C. The axis of the hole is represented by the unit vector $\hat{t}$. x, y, and z are a set of right-handed, mutually perpendicular axes*

[60] H. Poincaré, *J. Math., Paris*, [3], 1881, **7**, 375.
[61] H. Poincaré, *J. Math., Paris*, [3], 1882, **8**, 251.
[62] H. Poincaré, *J. Math., Paris*, [4], 1886, **2**, 151
[63] R B Fuller, U.S. P. 2 682 235, June 29, 1954.
[64] D. L. D. Caspar and A. Klug, *Cold Spring Harbor Symp. Quant. Biol.*, 1962, **27**, 1.
[65] R. B. Fuller, 'Utopia or Oblivion', Allen Lane, The Penguin Press, London, 1970, pp. 122, 123.
[66] W. F. Harris, ref. 1, p. 579.

Externally applied forces are removed and the body is allowed to relax. The result is a strained configuration, a distorted ring containing a right-handed screw dislocation lying along its axis. The translational displacement is the Burgers vector **b** of the dislocation (**b** ↑↑ **t̂**). Translation by **b** antiparallel to **t̂** (**b** ↑↓ **t̂**) results in a left-handed screw dislocation (Figure 7c). Relative translation of faces C$^+$ and C$^-$ perpendicular to **t̂** results in edge dislocations (Figure 7a, b). Volterra's *distorsions*[47] (Figure 7) of orders 1 and 2 are edge dislocations and the one of order 3 is a screw dislocation. Material is added where gaps are formed or removed where the body becomes self-penetrating.

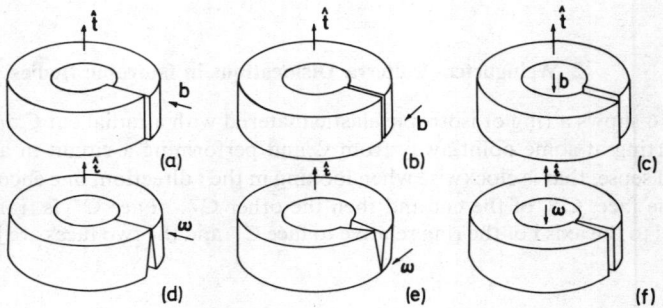

Figure 7 *Volterra's six elementary distorsions of orders 1 to 6. (a) and (b) represent edge dislocations: face C$^+$ of cut C in Figure 6 is translated by the Burgers vector **b** relative to face C$^-$: **b** ⊥ **t̂**. (c) represents a left-handed screw dislocation: **b** ↑↓ **t̂**. (d) and (e) are twist disclinations of rotation ω, where ω ⊥ **t̂** (cf. Figure 1). (f) is a negative wedge disclination, with ω ↑↓ **t̂** (cf. Figure 2). The vectors **b** are exaggerated*

Weingarten[45] demonstrated that the strain in the body is everywhere finite and twice differentiable if the relative displacement of the two faces of the cut is one possible for a rigid body. The location and shape of the cut and shape of the body do not matter provided only that the cut reduces the connectivity of the body without disconnecting it. Any second cut that can be continuously deformed into the first will result in the same state of strain in the body when its faces are subjected to the same relative rigid-body displacement.[47] The original cut leaves no trace in the strained body. The state of strain depends only on the geometry of the holes through the body, which when shrunk become singular lines, and the relative displacement.

Dispirations and Disclinations.—Any rigid-body displacement is a screw displacement: this is known as Chasles' Theorem.[67] Since defects arising from screw displacements are known as *dispirations*,[68] it follows that all

---

[67] J. L. Synge, *Handbuch der Physik*, 1960, **3** (1), 1.
[68] The word is formed from the Latin *spira* (a coil or twist) plus *di-* (in twain). A more complete discussion of the etymology of dispiration, dislocation, and disclination and of the verbal and adjectival forms is given in Appendix 2 of ref. 17.

defects of the Weingarten–Volterra type (including dislocations and disclinations) are dispirations in the widest sense of the term.[17,69] If the rotational component of the screw vanishes then the defect is a *dislocation*[70] (Figure 7a,b,c). If the translational component vanishes then the defect is a *disclination* (Figure 7d,e,f).

The rotational displacement associated with a disclination can be represented by a directed line segment ω of length proportional to the angle ω of rotation and lying along the axis of rotation. Its direction along the axis is that in which a right-handed screw would advance with the same rotation. ω and ω are both termed the *rotation* of the disclination: which one is meant is determined by the context. Here we adopt the convention that $\omega > 0$. In general a dispiration has associated with it both a Burgers vector **b** and a rotation ω. From Chasles' Theorem it follows that there are two categories of dispirations that are not either dislocations or disclinations: one has **b** ↑↑ ω, the other **b** ↑↓ ω. Suitable descriptive terms would be *co-dispirations* and *counter-dispirations*, respectively. In an isotropic continuum a dispiration can be viewed as a dislocation plus a disclination.

By analogy with *edge* (**b** ⊥ **t̂**) and *screw* (**b** ↑↑ ±**t̂**) dislocations, Nabarro[32] suggested the terms *edge* and *screw* disclinations for the special cases for which ω ⊥ **t̂** (Figure 7d,e) and ω ↑↑ ±**t̂** (Figure 7f), respectively. Others[9] preferred the more descriptive term *wedge* disclination for the latter. Since the conference on Fundamental Aspects of Dislocation Theory held in 1969 the terms *twist* and *wedge* disclination have predominated for the two cases, respectively.[71] The two special cases of the dispiration **b**, ω ⊥ **t̂** and **b**, ω ↑↑ ±**t̂** are termed *twist* or *edge* and *wedge* or *screw* dispirations, respectively.[17,69] Although wedge disclination and dispiration lines are frequently illustrated as coinciding with their rotations (Figures 2,4,5,7f) they are not necessarily coincident (Figure 8).

Wedge disclinations and dispirations are of two types: positive if ω ↑↑ **t̂** and negative if ω ↑↓ **t̂**. Negative wedge disclinations are frequently described as wedge disclinations with negative rotations, that is with rotations $-\omega$.

In an isotropic elastic body in which strains are everywhere finite, dispirations, including the special cases, dislocations and disclinations, can be viewed as lines threading holes in the body. The holes themselves represent cores

---

[69] W. F. Harris, *Phil. Mag.*, 1970, **22**, 949.

[70] *Dislocation* is commonly used in both wide and narrow senses. In metal physics it is often restricted to defects corresponding to Weingarten–Volterra defects of orders 1, 2, and 3 (Figure 7a—c). In continuum mechanics it includes orders 4, 5, and 6 (Figure 7d—f) as well, that is disclinations. It was introduced, in its wider sense, by A. E. H. Love, 'A Treatise on the Mathematical Theory of Elasticity', 3rd edn., Cambridge University Press, 1920, art. 156A.

[71] R. de Wit, ref. 1, p. 715. *Twist* and *wedge* are perhaps not quite as satisfactory as has been supposed. Just as a twist disclination can be made by inserting a wedge (Figure 1a) so can a wedge disclination be made with a twist. All published illustrations of twist disclinations in crystals other than Figure 1a, as far as the Reporter is aware, are similar to Figure 1b with an obvious twist. Had earlier illustrations been similar to Figure 1a with a clear wedge these terms might not have been adopted.

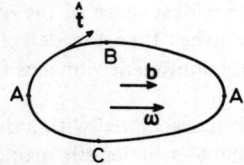

Figure 8 *A co-dispiration loop* (**b** ↑↑ ω). *At* A *it is a twist co-dispiration, at* B *a positive wedge co-dispiration and at* C *a negative wedge co-dispiration. At all other points it is a mixed co-dispiration*

of the defects. Without holes the defects can occur only as singular lines along which strains become infinite.

No dispiration can end within an elastic body except on another dispiration.[72]

Components of a General Dispiration.—Just as with dislocations, disclination and dispiration lines may be curved and **b** and ω do not vary along the lines. At different points along the line the defect may have twist, wedge, or mixed character (Figure 8).

In general a mixed dispiration line (in an isotropic body) at any point O on it can be decomposed into a dislocation plus a disclination. Each of these can be further decomposed so that any dispiration can be regarded as consisting of four components: edge and screw dislocations plus twist and wedge disclinations. There is no unique set of components. Provided $\omega < 2\pi$ rad one set can be obtained with the rotation ω_w of the wedge disclination component coincident with the tangent ($\pm \hat{\mathbf{t}}$) to the dispiration line at O and the axis of the rotation ω_t of the twist disclination passing through O: the disclination components are of the type shown in Figures 1a and 2. In some cases which remain to be fully explored the same is true for dispirations with $\omega \geqslant 2\pi$ rad.

A dispiration of rotation ω and Burgers vector **b** can be introduced into a ring like that in Figure 6 by rotating face C^+ of cut C through ω about ω and translating it by **b** (Figure 9). If the axis ω passes through O then any point P, on C^+, with position vector **p** relative to O is displaced to P' with position vector **p**' given by[73]

$$\mathbf{p}' = \mathbf{R} \cdot \mathbf{p} + \mathbf{b} \qquad (2)$$

where **R** is the *rotation tensor*, a *versor* in Gibbs' terminology:[73]

$$\mathbf{R} = \hat{\mathbf{a}}\hat{\mathbf{a}} + (\mathbf{I} - \hat{\mathbf{a}}\hat{\mathbf{a}}) \cos \omega + (\mathbf{I} \times \hat{\mathbf{a}}) \sin \omega \qquad (3)$$

---

[72] R. de Wit, ref. 1, p. 651.
[73] E. B. Wilson, 'Vector Analysis', Dover, New York, 1960; republication of the 2nd edn., Charles Scribner's Sons, 1909.

The Geometry of Disclinations in Crystals

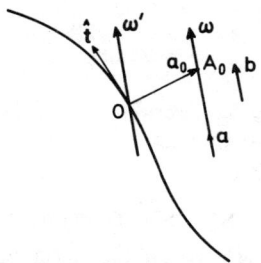

Figure 9 *A co-dispiration line with rotation ω, Burgers vector* **b** *and unit tangent* $\hat{\mathbf{t}}$ *at point O. Rotation ω has a direction given by the unit vector* $\hat{\mathbf{a}}$ *and passes through* A_0 *with position vector* $\mathbf{a}_0$ *relative to O*

where **I** is the unit isotropic tensor,

$$\mathbf{I} = \hat{\mathbf{i}}\hat{\mathbf{i}} + \hat{\mathbf{j}}\hat{\mathbf{j}} + \hat{\mathbf{k}}\hat{\mathbf{k}}$$

and $\hat{\mathbf{a}}$ is the unit vector in the direction of ω. In general ω does not pass through O but it may pass through a point A_0 with position vector $\mathbf{a}_0$. In that case

$$\mathbf{p}' - \mathbf{a}_0 = \mathbf{R} \cdot (\mathbf{p} - \mathbf{a}_0) + \mathbf{b} \tag{4}$$

This can be rearranged to give

$$\mathbf{p}' = \mathbf{R} \cdot [\mathbf{p} + \mathbf{a}_0 \cdot (\mathbf{R} - \mathbf{I}) + \mathbf{b}] \tag{5}$$

Comparison of equations (2) and (5) shows that the dispiration can be introduced by first introducing a dislocation with Burgers vector

$$\mathbf{b}' = \mathbf{a}_0 \cdot (\mathbf{R} - \mathbf{I}) + \mathbf{b} \tag{6}$$

and then a disclination of rotation ω', that is, rotation through angle ω about an axis through O in the $\hat{\mathbf{a}}$ direction. The disclination can be introduced first but then the displacement of the dislocation is

$$\mathbf{b}'' = \mathbf{R} \cdot \mathbf{b}' = (\mathbf{I} - \mathbf{R}) \cdot \mathbf{a}_0 + \mathbf{b} \tag{7}$$

Relative to a frame of reference attached to face C^- the Burgers vector depends on the order in which the two components are combined. Relative to a frame attached to face C^+ the Burgers vector is independent of the order. This is clear from equation (7).

The screw ($\mathbf{b}'_s$) and edge ($\mathbf{b}'_e$) components of the dislocation are simply

$$\mathbf{b}'_s = \mathbf{b}' \cdot \hat{\mathbf{t}}\hat{\mathbf{t}} \tag{8}$$

$$\mathbf{b}'_e = \mathbf{b}' - \mathbf{b}'_s = \mathbf{b}' \cdot (\mathbf{I} - \hat{\mathbf{t}}\hat{\mathbf{t}}) \tag{9}$$

The wedge (ω'_w) and twist (ω'_t) disclination components, with ω'_w and ω'_t both passing through O, are not in general obtained by equations of the same form as equations (8) and (9). For any rotation tensor **R** with rotation ω and axis $\hat{\mathbf{a}}$ one can define a vector **S** by

$$\mathbf{S} = \hat{\mathbf{a}} \tan \frac{\omega}{2} \tag{10}$$

It is Gibbs' *vector semi-tangent of version*.[73] The vector semi-tangents of the wedge and twist components then are

$$\mathbf{S}'_w = \mathbf{S}' \cdot \hat{\mathbf{t}}\hat{\mathbf{t}} \tag{11}$$

$$\mathbf{S}'_t = (\mathbf{S}' - \mathbf{S}'_w + \mathbf{S}' \times \mathbf{S}'_w)/(1 + \mathbf{S}' \cdot \mathbf{S}')$$
$$= \mathbf{S}' \cdot (\mathbf{I} - \hat{\mathbf{t}}\hat{\mathbf{t}} - \hat{\mathbf{t}}\hat{\mathbf{t}} \times \mathbf{S}')/[1 + (\mathbf{S}' \cdot \hat{\mathbf{t}})^2] \tag{12}$$

respectively. Equation (12) gives the rotation of the twist component relative to a frame of reference attached to face C^- if the twist component is introduced first. If the wedge component is introduced first then the twist component has a vector semi-tangent

$$\mathbf{S}''_t = \mathbf{R}_w \cdot \mathbf{S}'_t$$
$$= (\mathbf{S}' - \mathbf{S}'_w - \mathbf{S}' \times \mathbf{S}'_w)/(1 + \mathbf{S}'_w \cdot \mathbf{S}') \tag{13}$$

Relative to a frame of reference attached to C^+ the rotation of the twist component is independent of the order. For $\mathbf{S} \to \mathbf{0}$ equations (11) and (12) reduce to

$$\boldsymbol{\omega}'_w = \boldsymbol{\omega}'_t \cdot \hat{\mathbf{t}}\hat{\mathbf{t}} \tag{14}$$

$$\boldsymbol{\omega}'_t = \boldsymbol{\omega}' - \boldsymbol{\omega}'_w = \boldsymbol{\omega}' \cdot (\mathbf{I} - \hat{\mathbf{t}}\hat{\mathbf{t}}) \tag{15}$$

which have the same form as equations (8) and (9). From equation (11) the rotation ω'_w of the wedge component is given by

$$\tan \frac{\omega'_w}{2} = (\hat{\mathbf{a}} \cdot \hat{\mathbf{t}}) \tan \frac{\omega}{2} \tag{16}$$

Limitations on Rotations of Disclinations and Dispirations.—There is no published account of the limitations on the rotations of disclinations and dispirations in general. Here we mention a few special cases.

The positive wedge disclination of Figure 2 may be viewed as resulting from the removal of a sector of angle $\omega = \pi/2$ rad. Clearly a positive wedge disclination coincident with its rotation ω and with $\omega \geqslant 2\pi$ rad cannot occur in an elastic body in isolation. If ω lies outside the body on the other hand then there is no limit to ω in this case: an example with $\omega = 2\pi$ rad is shown in Figure 10. A positive wedge disclination with $\omega^+ \geqslant 2\pi$ and $\boldsymbol{\omega}^+$ within the crystal is possible if there is also present a parallel negative wedge disclination coincident with its rotation $\boldsymbol{\omega}^-$ and with $\omega^- > \omega^+ - 2\pi$ rad. A positive wedge disclination with $\omega = 2\pi$ rad and coincident with $\boldsymbol{\omega}$ is possible in conjunction with a dislocation (Figure 11).[17,66] If the crystal is slit along a radius it relaxes and $\mathbf{b}_0$ becomes $\mathbf{b}$ the Burgers vector of the dislocation. The presence of the dislocation is indicated by the terminating radial planes.

The Geometry of Disclinations in Crystals

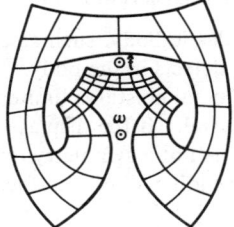

Figure 10 *A positive wedge disclination of rotation 2π rad. The rotation ω is parallel to the unit tangent $\hat{t}$ but not coincident with the disclination line. ω lies outside the body. It is difficult to imagine any physical system in which such a defect could exist*

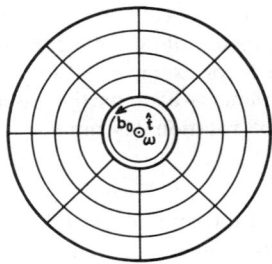

Figure 11 *A positive wedge disclination with rotation $\omega = 2\pi$ rad and coincident with ω. The disclination has associated with itself an edge dislocation with Burgers vector $\mathbf{b}$. If the crystal is slit along a radius it will relax becoming rectangular in cross-section: the curved arrow $\mathbf{b}_0$ will become the Burgers vector $\mathbf{b}$ of the dislocation. The dislocation may be viewed as eight elementary dislocations each indicated by the termination of a radial lattice plane. This type of disclination was first described by Nabarro.[74] The lattice 'planes' may be spiral in cross-section instead of straight or circular* (cf. *Figure* 3d,e, *and* f).

There is no limit to the rotation of negative wedge disclinations.

The twist disclination of Figure 1a intersects its rotation ω: it can be made by inserting a sector of angle $\omega = \pi/2$ rad. There is no limit to the magnitude of ω. On the other hand, if ω is reversed, sectors of angle ω are removed and ω is limited to values less than 2π rad. It follows that an isolated twist disclination with $\omega \geqslant 2\pi$ rad cannot intersect its rotation ω within an elastic body. There is no limit to ω for twist disclinations of the type shown in Figure 1b which do not intersect their rotations. Nabarro illustrates such a twist disclination with $\omega = 2\pi$ rad.[75]

The rotation of dispirations is limited in the same way as the rotation of disclinations.

---

[74] Ref. 4, Figure 3.3, section 3.1.1.
[75] Ref. 4, Figure 3.2.

3 Disclinations and Dispirations in Structured Materials

Texture or structure in an elastic body, such as a crystal, limits the possible displacements (**b** and **ω**) for defects in the body.[76] For example, the Burgers vector of a dislocation in a metal crystal is limited to translational symmetry operations of the perfect crystal. If the displacement **b**, like those displacements in Figure 7a,b, and c, is not a symmetry operation then the lattice will not match across the join: the join will be a surface defect which terminates at a partial dislocation.

Corresponding to the three proper symmetry operations of crystals, translational, rotational, and screw symmetry, there are three distinct types of Volterra–Weingarten defect: dislocations, disclinations, and dispirations.

Disclinations in Crystals.—The rotation ω of a disclination is a rotational symmetry operation of the perfect crystal.[77,78] Fourfold axes of rotational symmetry in the perfect crystal allow the wedge and twist disclinations of rotation $\pi/2$ rad in Figure 1, for example. If **ω** is not a symmetry operation then a surface defect terminates at a partial disclination.[78] In general a crystal with s-fold axes of rotational symmetry can have disclinations of rotation[77]

$$\omega = 2\pi p/s \tag{17}$$

only,[78] where p is a positive integer and $s = 1, 2, 3, 4$, or 6. It follows that the smallest possible disclination in a crystal has $\omega = \pi/3$ rad.[78] A positive wedge disclination coincident with its rotation **ω** and with $\omega = \pi/3$ rad is shown in Figure 12. Wedge disclinations coincident with their rotations and with integer p destroy the s-fold symmetry of the perfect crystals and become axes of $(s \mp p)$-fold symmetry: the negative and positive signs apply to positive and negative wedge disclinations, respectively. This is seen in Figures 2—5, 12—14, and 19.

A useful method for investigating disclinations in a crystal was devised by Nabarro.[77] One constructs a closed curve in the crystal but instead of counting steps, as one does in a Burgers circuit around a dislocation, one observes the change in local orientation of the lattice along the curve. Figure 13 shows a positive wedge disclination of rotation $\pi/2$ rad encircled by a circuit. Starting at point P and passing once around the circuit in a clockwise or anticlockwise sense one finds the lattice rotates through an angle ω' in the same sense about an axis parallel to the rotation of the disclination. If the strain in the neighbourhood of P is relaxed then ω' becomes $\pi/2$ rad, the rotation of the disclination. For the negative wedge disclination of Figure 14 the

---

[76] Ref. 38, p. 5.
[77] Ref. 4, section 3.1.1.
[78] The only conceivable exception is for a small twist disclination loop with rotation ω along the axis of the loop. For example the loop may encircle one fibre in a crystalline bundle of fibres each of which when isolated has an axis of ms-fold rotational symmetry where m is an integer >1. In that case $\omega = 2\pi p/ms$ rad.

The Geometry of Disclinations in Crystals

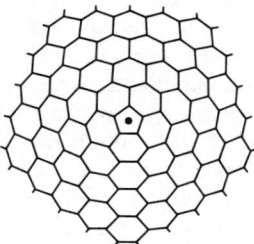

Figure 12 *A positive wedge disclination of rotation $\pi/3$ rad. The defect has unit tangent $\hat{t}$ normal to the Figure and coincides with its rotation ω at the centre. The perfect crystal has sixfold axes of symmetry. The disclinated crystal has one fivefold axis coinciding with the disclination*

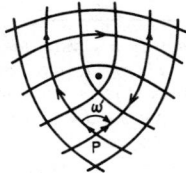

Figure 13 *A positive wedge disclination line of rotation $\pi/2$ rad normal to the plane of the Figure. The line coincides with its rotation. The lattice rotates clockwise through an angle ω' when one passes clockwise around a closed circuit encircling the defect. In strain-free crystals ω' becomes the rotation $\pi/2$ rad*

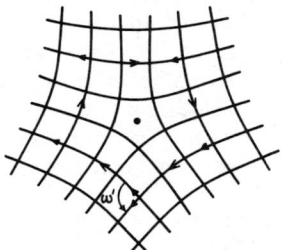

Figure 14 *A negative wedge disclination of rotation $\pi/2$ coincident with but antiparallel to ω. The lattice rotates anticlockwise through ω' along a clockwise circuit. In strain-free crystals ω' becomes $\pi/2$ rad*

lattice rotates in a sense opposite to that of the directed circuit. Similarly for the disclinations of rotation 2π rad in Figures 10 and 11 the lattice rotates about an axis parallel to the disclination line through an angle 2π rad. For the twist disclinations of Figure 1 the lattice rotates through an angle $\pi/2$ rad, measured in a strain-free crystal, about an axis perpendicular to the defect line when one passes once around a circuit encircling the line. The angle ω' in the strained crystal may be termed the *local rotation*[17] of the disclination and is the analogue of the local Burgers vector in a dislocated crystal.[79]

The only type of disclination that can be expected in a solid crystal, Friedel and Kléman[7] argue, is a straight wedge disclination coincident with its rotation (*e.g.* Figures 2 and 12—14): all other configurations would involve prohibitively high strain energies. In crystalline polymers and other bodies built up of regularly arranged chains or fibres small twist disclination loops, each encircling one fibre, are also possible[15] (Figure 15).

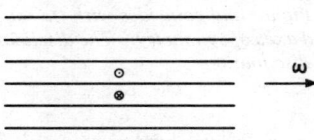

Figure 15 *A twist dispiration loop in a crystalline polymer or crystalline array of fibres. The loop encircles one chain the axis of which is the rotation ω. The dispiration is of a similar type to the one in Figure 16b. If the Burgers vector vanishes then the dispiration becomes a disclination loop, the disclination being of a similar type to the one in Figure 1b*

The strain energy and strain field of disclinations and many other problems have been reported in numerous papers.[11, 15, 72, 80]

Dispirations in Crystals.—If the material of the body in Figure 6 is crystalline then a dispiration can be introduced only if the relative displacement suffered by the two faces of the cut C is a screw symmetry operation of the crystal. The translational, and rotational components, **b** and ω, of the dispiration are not in general symmetry operations. If they are not symmetry operations then the translation **b** and the rotation ω, each performed on its own, would result respectively not in a dislocation and a disclination but a *partial* dislocation and a *partial* disclination. They are, however, partial defects *without* attached surface defects. This is what makes the dispiration a distinct crystal

---

[79] B. A. Bilby, R. Bullough, and E. Smith, *Proc. Roy. Soc.*, 1955, **A231**, 263; B. A. Bilby and E. Smith, *ibid.*, 1956, **A236**, 481.
[80] K. H. Anthony, ref. 1, p. 637; *Arch. Rat. Mech. Anal.*, 1970, **37**, 161; 1971, **40**, 50; W. Huang and T. Mura, *J. Appl. Phys.*, 1970, **41**, 5175; 1972, **43**, 240; G. C. T. Liu and J. C. M. Li, *ibid.*, 1971, **42**, 3313; T.-W. Chou, *ibid.*, 1971, **42**, 4092, 4931; E. S. P. Das and M. J. Marcinkowski, *ibid.*, 1971, **42**, 4107; H. H. Kuo and T. Mura, *ibid.*, 1972, **43**, 1454, 3936; T. Mura, *Arch. Mech. Stos.*, 1972, **24**, 449; T.-W.Chou and T.-L. Lu, *J. Appl. Phys.*, 1972, **43**, 2562; F. Kroupa and L. Lejček, *Phys. Status Solidi (B)*, 1972, **51**, K121; J. C. M. Li, *Surface Sci.*, 1972, **31**, 12; in 'Nature and Behavior of Grain Boundaries', ed. H. Hu, Plenum, New York, 1972, p. 71; J. Dunders, *Phys. Status Solidi (B)*, 1972, **53**, 157; T.-W. Chou and Y. C. Pan, *J. Appl. Phys.*, 1973, **44**, 63; R. de Wit, *Arch. Mech. Stos.*, 1972, **24**, 499; *J. Res. Nat. Bur. Stand. (A)*, 1973, **77**, 49, 359; 1973, **77**, 607; *Phys. Status Solidi (A)*, 1973, **18**, 669.

defect and not simply a dislocation plus a disclination. The surface defects associated with the two partial defects may be viewed as mutually cancelling. The earliest suggestion that there might be defects corresponding to screw symmetry appears to be Friedel's[81] although the first illustrations of such defects were published much later.[17,69]

The rotational component of the dispiration in a crystal with s-fold screw axes of symmetry is restricted to values given by equation (17).[17,69,82] For a screw axis of symmetry s_m the translational and rotational components of the displacement are each symmetry operations if n is a multiple of s^0, where s^0 is s divided by the highest common factor of the integers s and m. Only in that case can the dispiration be regarded as a disclination plus a dislocation. As with the disclination the smallest value that the rotation ω of a dispiration can have is $\pi/3$ rad.[82] Again as with the disclination the lattice rotates through an angle ω' along a closed curve encircling the dispiration; measured in strain-free crystal ω' becomes the rotational component ω of the dispiration.

Corresponding to the twist disclinations of Figure 1 there are the twist dispirations of Figure 16, and to the positive wedge disclination of Figure 2 the positive wedge dispiration of Figure 17. The dispirations of Figures 16 and 17 have $\omega = \pi/2$ rad and are possible because of 4_3 (Figure 16a) and 4_1 (Figures 16b and 17) screw axes of symmetry in the perfect crystal.

In a solid crystal the only dispirations likely are straight wedge dispirations coincident with their rotations and in crystalline polymers and arrays of fibres small twist dispiration loops encircling a fibre and the axis of rotation.[69] Reneker's[83] 'point dislocations' in crystalline polymers are twist dispiration loops.

Defects in Surface Crystals.—Disclinations and dispirations are more likely in bodies of limited extent in one dimension.[3] An extreme case is the surface crystal.[17,66] A perfect surface crystal has units arranged on a two-dimensional lattice. The crystal itself has thickness and is preferably not described as two-dimensional. It is most common perhaps in biology, appearing in many membranous forms such as some cell walls and membranes and the capsids of viruses.[17]

At first sight it might be supposed that the types of disclinations and other defects possible in surface crystals might be more limited than in conventional crystals with translational periodicity in three dimensions. Indeed the only obvious defects of the Weingarten–Volterra type possible would seem to be the edge dislocation and the wedge disclination. These are the only defects that can be illustrated in a plane drawing with no depth cues. Surprisingly, perhaps, the types of defects possible in surface crystals are more

---
[81] Ref. 38, p. 6.
[82] A conceivable exception is a small twist dispiration loop. *Cf.* footnote 78.
[83] D. H. Reneker, *J. Polymer Sci.*, 1962, **59**, S39.

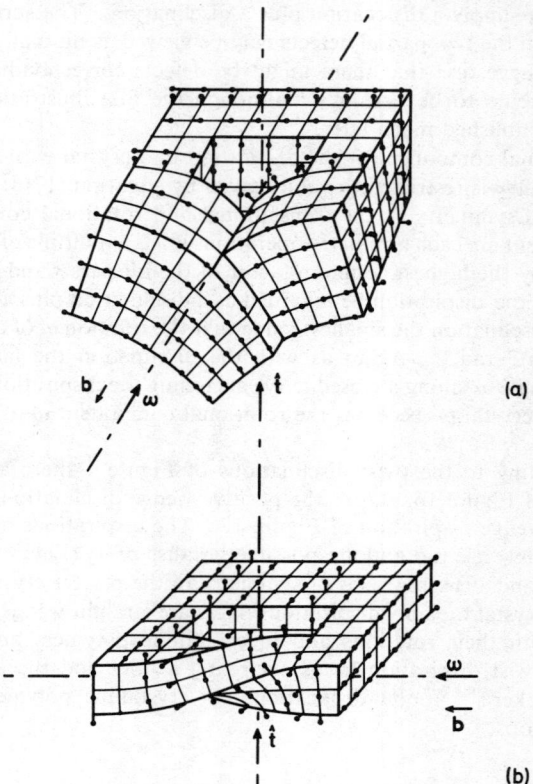

Figure 16 *Twist dispirations with rotation $\pi/2$ rad. The rotation $\boldsymbol{\omega}$ is perpendicular to the dispiration line with unit tangent $\hat{\mathbf{t}}$.* (a) *is a twist counter-dispiration* ($\mathbf{b}\uparrow\downarrow\boldsymbol{\omega}$) *which intersects its rotation; the crystal has 4_3 screw axes of symmetry.* (b) *is a twist co-dispiration* ($\mathbf{b}\uparrow\uparrow\boldsymbol{\omega}$) *which does not intersect its rotation; the crystal has 4_1 screw axes of symmetry*

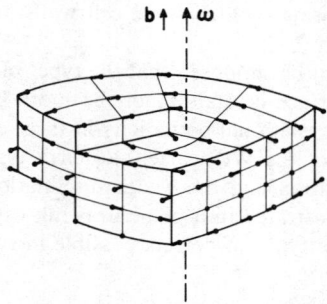

Figure 17 *A positive wedge co-dispiration of rotation $\pi/2$ rad in a crystal with 4_1 screw axes of symmetry. $\boldsymbol{\omega}$ is parallel to $\hat{\mathbf{t}}$ and coincident with the dispiration line*

varied than in conventional crystals. It is only the edge dislocation and the wedge disclination that are strictly analogous to the defects in conventional crystals.

Starting with a thin circular sheet of perfect surface crystal with a co-axial hole, instead of the body of Figure 6, one introduces a radial cut and displaces one face of the cut relative to the other by a symmetry operation of the perfect crystal. The only symmetry properties of interest are: (i) translations in the plane of the lattice; (ii) axes of s-fold rotational symmetry ($s = 1,2,3,4,$ or 6) normal to the plane of the lattice; (iii) onefold and twofold axes of rotational symmetry in the mid-plane of the crystal; (iv) 2_1 screw axes in the mid-plane; and (v) onefold axes neither normal to nor within the mid-plane of the crystal.[17] Operations (iii), (iv), and (v) imply displacements out of the plane of the surface lattice. These have no analogue in conventional crystals in which all symmetry operations imply displacements within the space of the crystal. Operations (i) and (ii) lead to *intrinsic* defects in surface crystals while the remaining operations lead to *extrinsic* defects.[17,66] In this sense all defects in conventional crystals are intrinsic.

The translations of operation (i) lead to intrinsic edge dislocations, there being no analogue in surface crystals of screw dislocations. The dislocations in the surface crystal formed in Bragg's bubble raft[84-86] are good examples.

Intrinsic Disclinations. The rotations of operation (ii), with $s = 2,3,4,$ or 6, lead to intrinsic wedge disclinations like those of Figures 12—14. Positive wedge disclinations of rotation $\pi/3$ rad can be introduced into an array of bubbles[87-90] (Figure 18). Freed of externally applied forces rubber-ball and paper models of surface crystals containing wedge disclinations deform out of the plane. Positive wedge disclinations coincident with their rotations become circular cones with the disclination at the apex. The apical angle α, that is the angle between the axis of the cone and a generator, is given by

$$\sin \alpha = \frac{2\pi - \omega}{2\pi} \qquad (18)$$

where $0 < \omega < 2\pi$ rad.

A rubber-ball model with $\omega = \pi/3$ rad is shown in Figure 19: paper models with other rotations $< 2\pi$ rad are shown elsewhere.[17,91] The case with $\omega = \pi/3$ rad is common in biological surface crystals occurring in the cell walls

---

[84] W. L. Bragg and J. F. Nye, *Proc. Roy. Soc.*, 1947, **A190**, 474; W. L. Bragg and W. M. Lomer, *ibid.*, 1949, **A196**, 171.
[85] W. M. Lomer, *Proc. Roy. Soc.*, 1949, **A196**, 182.
[86] W. L. Bragg, W. M. Lomer, and J. F. Nye, 'Experiments with Bubble Model of a Metal Structure', motion picture film produced by N. S. Macqueen.
[87] W. F. Harris, unpublished observations.
[88] Y. Ishida, *Seisankenkyu*, 1972, **24**, 477.
[89] Y. Ishida, *Kinzoku*, 1973, **43** (6), 64.
[90] Y. Ishida, private communication.
[91] H. Steinhaus, 'Mathematical Snapshots', Oxford University Press, New York, 1960, pp. 244–247.

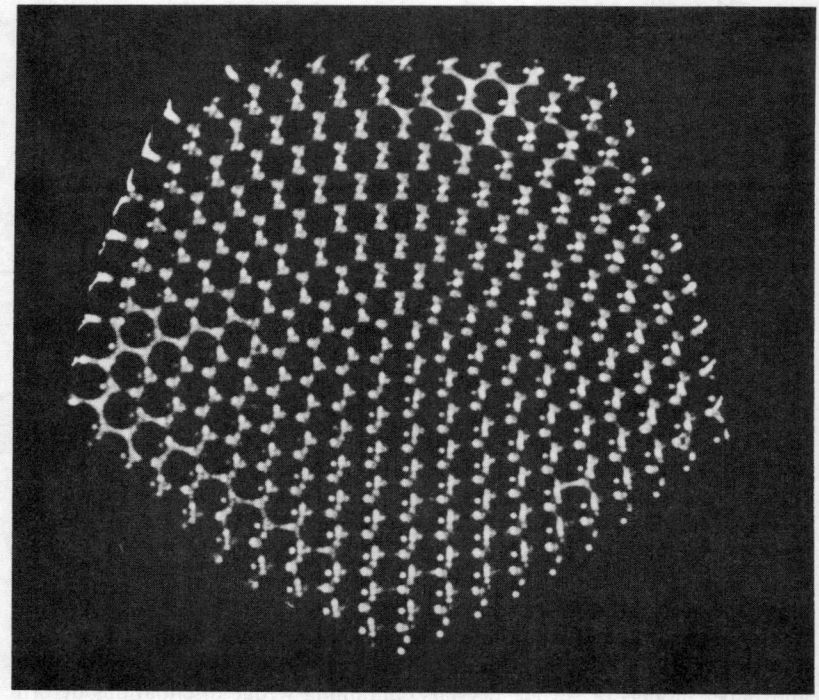

Figure 18 *Positive wedge disclination of rotation* $\pi/3$ *rad in an array of bubbles* (cf. *Figure* 12). *The diameter of the bubbles is* 2 mm. (*Photograph supplied by Y. Ishida*)

of some bacteria[92,93] (Figure 20) and in the protein capsids of spherical viruses[17,66] (Figure 21). Examples of viruses showing positive wedge disclinations of rotation $\pi/3$ rad are found in the work of Caspar, Klug, and others.[64,94-99] Robertson[100] published photographs of models of plasma

[92] W. F. Harris and L. E. Scriven, *Nature*, 1970, **228**, 827.
[93] F. R. N. Nabarro and W. F. Harris, *Nature*, 1971, **232**, 423.
[94] E. Kellenberger, *Scientific American*, 1966, **215** (6), 32.
[95] E. Kellenberger, in 'Principles of Biomolecular Organization' (Ciba Foundation Symp.) ed. G. E. W. Wolstenholme and M. O'Connor, Churchill, London, 1966, p. 192.
[96] R. Hull, G. J. Hills, and R. Markham, *Virology*, 1969, **37**, 416.
[97] N. A. Kiselev and A. Klug, *J. Mol. Biol.*, 1969, **40**, 155.
[98] N. G. Wrigley, *J. Gen. Virol.*, 1969, **5**, 123; Ph.D. Thesis, Australian National University, Canberra, 1969; W. G. Laver, H. Banfield Younghusband, and N. G. Wrigley, *Virology*, 1971, **45**, 598.
[99] J. T. Stasny, A. R. Neurath, and B. A. Rubin, *J. Virol.*, 1968, **2**, 1429.
[100] J. D. Robertson, in 'Handbook of Molecular Cytology', ed. A. Lima-de-Faria, North Holland, Amsterdam and London, 1969, p. 1403.

The Geometry of Disclinations in Crystals

Figure 19 *Rubber-ball model of a surface crystal containing a positive wedge disclination of rotation $\pi/3$ rad. The model is conical with the disclination located at the apex*

membranes containing these defects although there is as yet no clear evidence of the existence of disclinations in biological membranes. The theoretical problem of buckling of thin plates containing wedge disclinations has been considered by Mitchell and Head.[101]

The areolae in the diatoms *Thalassiosira eccentrica* and *T. symmetrica*, described by Fryxell and Hasle,[102] form a surface crystal at the centre of which is located a negative wedge disclination of rotation $\pi/3$ rad[103] (Figure 22). Models,[17] made out of paper, of negative wedge disclinations coincident with their rotations are shown in Figure 23 ($\omega = \pi/2$ rad and 2π rad in a and b, respectively). Formally the surfaces are cones.

Intrinsic dispirations are not possible nor are intrinsic twist disclinations.[17]

An important positive wedge disclination in biological surface crystals has rotation 2π rad and has associated with it a dislocation (Figure 11 viewed

---

[101] L. H. Mitchell and A. K. Head, *J. Mech. Phys. Solids*, 1961, **9**, 131.
[102] G. A. Fryxell and G. R. Hasle, *J. Phycol.*, 1972, **8**, 297.
[103] W. F. Harris and G. A. Fryxell, *Nature*, to be submitted.

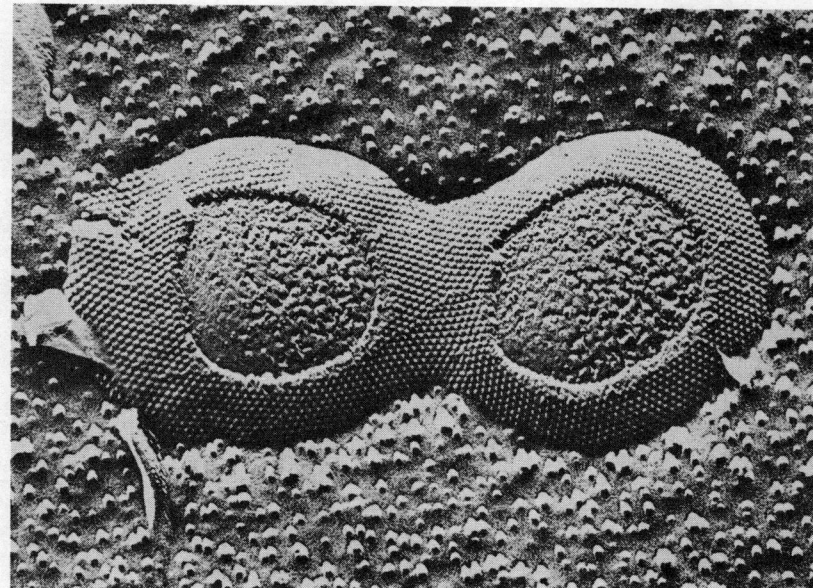

Figure 20 Wedge disclinations in the outer cell wall of a dividing cell of the marine nitrifying bacterium (Nitrosomonas, sp.). ($\times$ 58 200.) Specimen prepared by freeze-etching. The wall is composed of subunits, approximately 15 nm in diameter, in a hexagonal array. The wall can be viewed as two fused truncated cones (cf. Figure 19). A positive wedge disclination of rotation $\pi/3$ rad is situated at the apex of the right-hand cone. The disclination of the left-hand cone occurs as two partial disclinations with a total rotation of $\pi/3$ rad and which are connected by a grain boundary. Four edge dislocations can be identified
(Reproduced by permission from Science, 1969, **163**, cover of issue no. 3868 Copyright 1969 by the American Association for the Advancement of Science)

as a surface crystal). If the structure is thin in the direction normal to the figure it can release much of the strain by deforming into a tube-shaped structure termed a *cylindrical crystal*[17,104] (Figure 24). The cylindrical crystal may be viewed as a limiting case of a conical crystal, with apical angle α equal to zero [cf. equation (18)]. The Burgers vector of the dislocation associated with the disclination gives the circumference of the cylindrical crystal; it is useful for characterizing cylindrical crystals and is termed the *characteristic vector*.[104] Cylindrical crystals are common in biology and occur as bacterial flagella,[105] the capsids of rod-shaped viruses,[106] microtubules,[107] and other structures.[17]

[104] W. F. Harris and L. E. Scriven, *J. Theoret. Biol.*, 1970, **27**, 233.
[105] W. F. Harris and L. E. Scriven, *J. Mechanochem. Cell Motility*, 1971, **1**, 33.
[106] W. F. Harris, *Nature*, 1972, **240**, 294.
[107] W. F. Harris, *J. Mechanochem. Cell Motility*, 1972, **1**, 147.

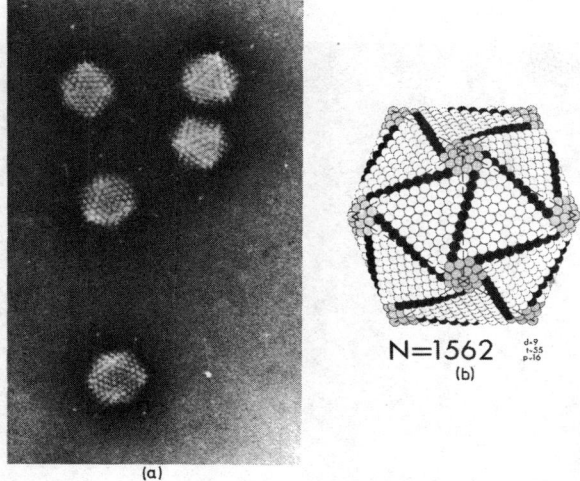

Figure 21 *Spherical viruses.* (a) *An electron micrograph of chick embryo lethal orphan virus.* (× 100 000.) (b) *A model of* Sericesthis *iridescent virus showing some of the positive wedge disclinations of rotation* $\pi/3$ *rad located at the vertices of the icosahedral shell*
[Reproduced by permission from (a) *Virology,* 1971, **45**, 598 and (b) N. G. Wrigley, Ph. D. Thesis, Australian National University, Canberra, 1969]

Disclinations, and dispirations, in conventional crystals appear as line singularities in the crystal lattice. In surface crystals, however, one sees that disclinations are not necessarily associated with singularities. Nowhere in the lattice of the disclinated crystal of Figure 24 is there a singularity. If the defect is associated with a singularity then it is termed a *local* defect; if not then it is a *global* defect.[17,66] In conventional crystals all defects are local. In surface crystals local defects are point singularities.

The disclination of Figure 24 can also be introduced by an extrinsic operation, that is by rotation about onefold axes of type (v). The term extrinsic defect is better restricted to those defects that cannot be made intrinsically.

Extrinsic Disclinations and Dispirations. The simplest extrinsic defect is a twist disclination of rotation π rad resulting from operation (iii).[17] An example is shown in Figure 25. The corresponding twist dispiration is similar[17] (Figure 26). These two defects are the equivalent in surface crystals of the defects in conventional crystals shown in Figures 1b and 16b, respectively. There are no equivalents of the total twist disclination and dispiration of Figures 1a and 16a: twist disclinations and dispirations in surface crystals cannot lie on their rotations. The disclinated and dispirated crystals of Figures 25 and 26 have the topology of a Möbius strip of one half-turn. In general, twist disclinations and dispirations of rotation $n\pi$ rad in surface

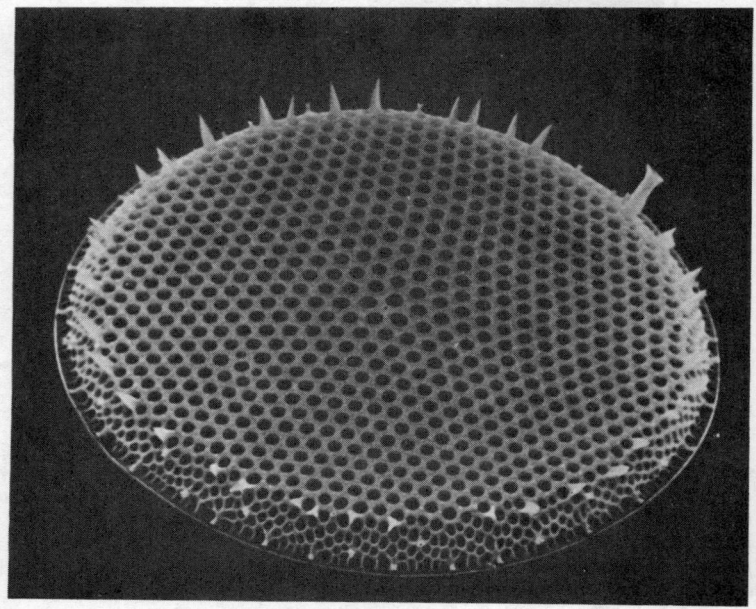

Figure 22 *Negative wedge disclination of rotation* $\pi/3$ *rad in the pattern of areolae in the diatom* Thalassiosira eccentrica. *The defect coincides with its rotation and is located at the centre of the diatom* ($\times$ 1980)
(Reproduced by permission from *J. Phycol.*, 1972, **8**, 297)

crystals result in Möbius strips of n half-turns. If the rotation $n\pi$ rad of an extrinsic defect has n odd then the surface is one-sided.

Extrinsic and intrinsic operations can be combined in many ways.[17] Two of the simpler examples are shown in Figures 27 and 28. Other interesting examples are found in the graphic work of the Dutch artist M. C. Escher: his 'Swans' (Figure 29) may be viewed as containing a pair of defects each of which consists of a positive wedge disclination ($\omega = 3\pi/2$ rad) plus a twist disclination ($\omega = \pi$ rad) while each of the two defects in his 'Horseman' (Figure 29) may be viewed as wedge dispirations ($\omega = \pi$ rad). The reporter knows of no examples of extrinsic defects in natural surface crystals.

Closed Surface Crystals. A surface crystal can close to form a shell. For example the ends of a cylindrical crystal (Figure 24) can be joined to make a toroidal crystal containing no local defects. The crystal contains two positive wedge disclinations of rotation 2π rad each in association with a dislocation.[17,66] Figure 21 shows examples of closed (spherical) surface crystals containing local positive wedge disclinations of rotation $\pi/3$ rad.

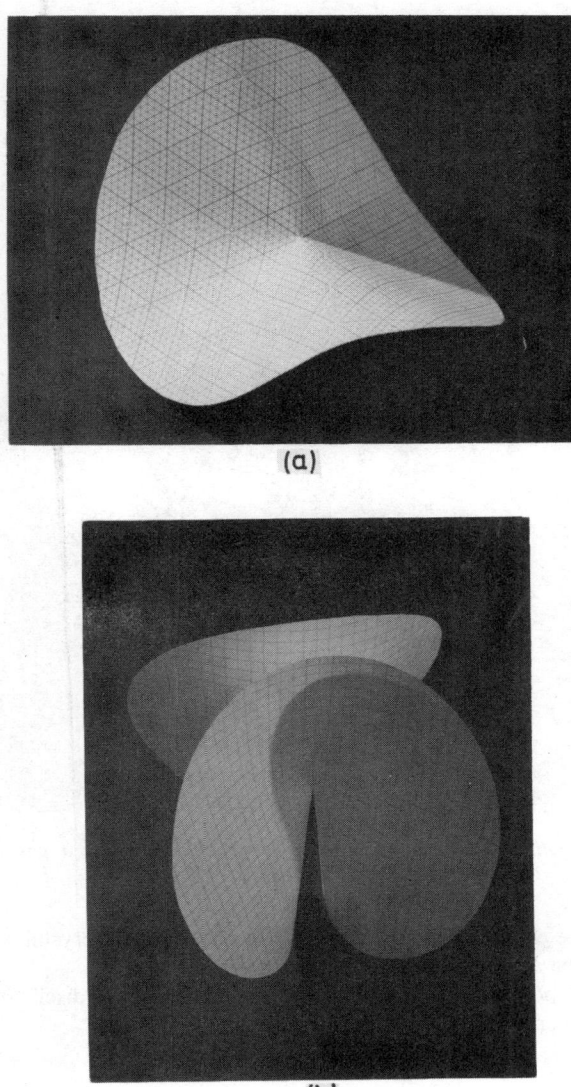

Figure 23 *Paper models of negative wedge disclinations of rotations $\pi/3$ rad (a) and 2π rad (b) in a surface crystal. The disclinations coincide with their rotations*

Figure 24 *A cylindrical surface crystal. The crystal contains a positive wedge disclination of rotation 2π rad lying along its axis*

Closed surface crystals have no analogue in conventional crystals: the analogue would be a crystal with no bounding surface.

There is a topological restriction on the rotations of local disclinations in a closed surface crystal:[17,66]

$$\sum_i \omega_i - \sum_j \omega_j = 2\pi\chi \tag{19}$$

where the ω_i and ω_j are the rotations of the positive and negative wedge disclinations, respectively, and χ is the Euler–Poincaré characteristic of the closed surface. For a surface topologically equivalent to a sphere $\chi = 2$, while for a topological torus $\chi = 0$. It follows that spherical surface crystals

The Geometry of Disclinations in Crystals 83

Figure 25 *An extrinsic twist disclination of rotation π rad in a surface crystal. The surface has the topology of a Möbius strip of one half-turn;* (cf. *Figure* 1b)

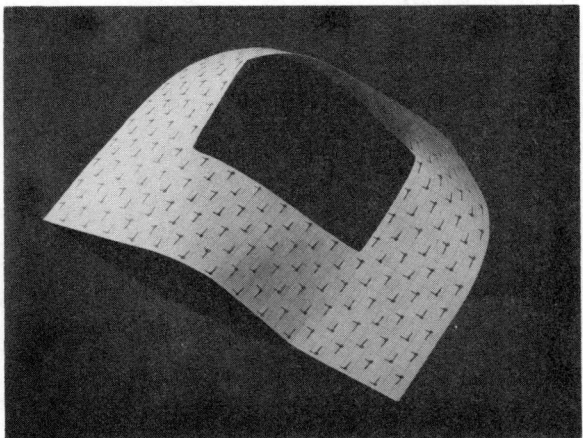

Figure 26 *An extrinsic twist dispiration with rotation π rad in a surface crystal;* (cf. *Figures* 16b *and* 25). *The perfect surface crystal does not have twofold axes of rotational symmetry from which it follows that a twist disclination like that of Figure* 25 *cannot be introduced in spite of the similarity of the two defects*

necessarily contain local wedge disclinations while it is possible to have toroidal crystals without them. For a spherical surface crystal with local wedge disclinations all of the same rotation one has the following possibilities only: (i) twelve wedge disclinations of rotation $\pi/3$ rad; (ii) eight of rotation $\pi/2$ rad; (iii) six of rotation $2\pi/3$ rad; (iv) four of rotation π rad;

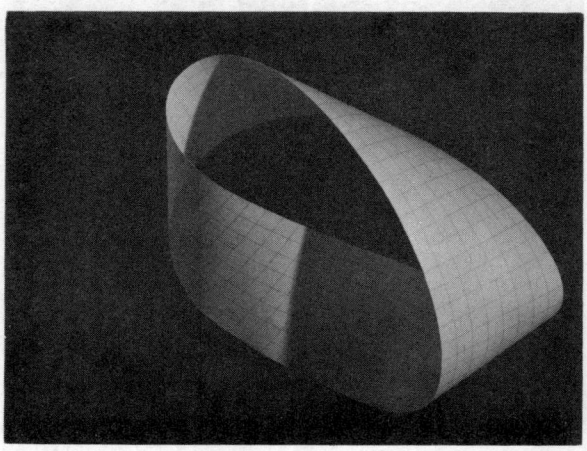

Figure 27 *A surface crystal containing a positive intrinsic wedge disclination of rotation* 2π *rad plus a dislocation* (cf. *Figure* 24) *together with a twist disclination of rotation* π *rad* (cf. *Figure* 25)

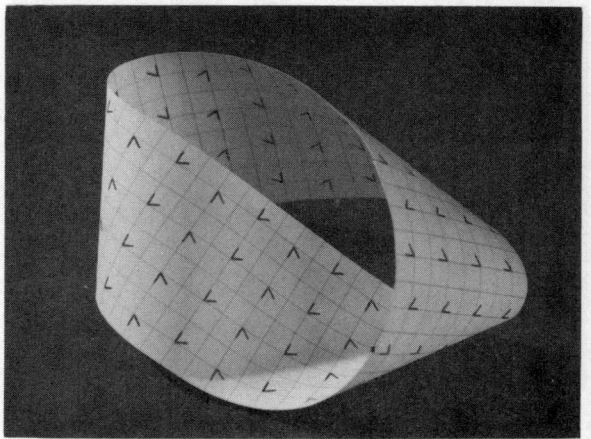

Figure 28 *A surface crystal containing a positive intrinsic wedge disclination of rotation* 2π *rad plus a dislocation together with a twist dispiration of rotation* π *rad* (cf. *Figures* 26 *and* 27)

(v) three of rotation $4\pi/3$ rad; (vi) two of rotation 2π rad; (vii) one of rotation 4π rad. The first, (i), is common, occurring in the capsids of spherical viruses (Figure 21). In these viruses the disclinations coincide with their rotations: they are distributed uniformly over the surface crystal so that they

The Geometry of Disclinations in Crystals

Figure 29 'Swans' *and* 'Horseman' *by the Dutch artist* M. C. Escher. *Each of the two defects in* 'Swans' *may be viewed as consisting of a positive wedge disclination of rotation* $3\pi/2$ rad *plus a twist disclination of rotation* π rad *while each in* 'Horseman' *is a type of wedge dispiration of rotation* π rad
(Reproduced by permission of the Escher Foundation, Haags Gemeentemuseum, The Hague)

lie at the corners of an icosahedral shell. In a similar fashion (ii), (iii), and (iv) could be represented by disclinations located at the corners of cubic, octahedral, and tetrahedral shells respectively. The only other Platonic body, the dodecahedron, would require 20 wedge disclinations each of rotation $\pi/5$

rad which is impossible in a surface crystal because of the crystallographic restriction: a dodecahedral shell, therefore, cannot be a disclinated surface crystal. Possibility (v) leads to a shell which would appear triangular in a section containing the three disclinations and lenticular in any section normal to the first. Each of the two disclinations of (vi) would lie at the ends of a spindle-shaped or rod-shaped shell: dislocations must be present as well. Possibility (vii) would represent an unlikely shell and is not discussed further.

Disclinations in Other Materials.—In the past six years a large number of papers has appeared on disclinations in liquid crystals. We discuss here some of the characteristics of disclinations in these materials primarily for purposes of comparison with crystalline bodies. For more complete accounts the reader is referred to several papers.[7,12,37,108—113]

If a defect is introduced into a liquid crystal by the process described in Section 2 above, then in addition to elastic relaxation there may be plastic relaxation.[7] A liquid crystal has symmetry elements which are not quantized. Defects with these elements as Burgers vectors or rotations tend to be dissipated by plastic relaxation. Thus it is only those defects with Burgers vectors and rotations which are quantized symmetry elements that are stable.

Apart from onefold axes the only quantized symmetry operations of nematic liquid crystals (Figure 30a) are twofold axes normal to the molecules. Thus the only Volterra–Weingarten defects likely in a nematic at rest are disclinations of rotation $n\pi$ rad: this includes the wedge disclinations of Figure 3, twist disclinations (Figure 30b) and disclinations of mixed charac-

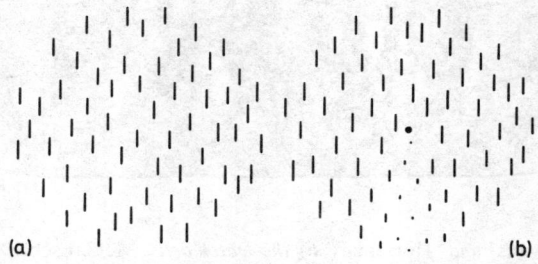

Figure 30 (a) *A perfect nematic liquid crystal.* (b) *A nematic containing a twist disclination line ($\omega = \pi$ rad) normal to the Figure. Crossing the lower half of (b) one finds the orientation of the molecules, initially parallel to the plane of the Figure, twisting through π rad about an axis in the plane of the Figure*

[108] Groupe Expérimental d'Etudes de Cristaux Liquides, *J. Phys.*, *Paris*, 1969, **30**, C4-38.
[109] J. Rault, Thèse, Université Paris-Sud, Centre d'Orsay, 1972.
[110] J. Rault, *Phil. Mag.*, 1973, **28**, 11.
[111] Y. Bouligand, *J. Phys.*, *Paris*, 1972, **33**, 715.
[112] Y. Bouligand, in 'Dislocation Theory—A Treatise', ed. F. R. N. Nabarro, in press.
[113] M. Kléman, *Bull. Soc. franç. Min. Crist.*, 1972, **95**, 215.

ter.[7,114] Wedge disclination lines coincide with their rotations. The cores of twist disclinations are not large and hollow as they are in crystals (*cf.* Figure 1).

In the case of wedge disclinations of rotation $n\pi$ rad where n is even, further plastic relaxation is possible near the disclination line. The molecules near the core tend to become parallel to the line and the singularity is removed.[115] If the line intersects a bounding surface which constrains the molecules near it to remain parallel to it then that point of intersection is all that remains of the singularity (Figure 31). Further discussion of these

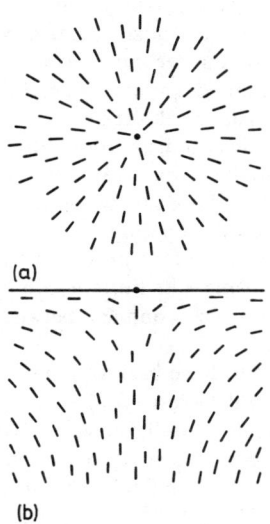

Figure 31 (a) *A wedge disclination line of rotation 2π rad in a nematic seen end on* (cf. *Figure* 3e). (b) *A side view of the same defect showing the molecules tending to become parallel to the defect and thus removing the line singularity. A point singularity remains at the surface of the liquid crystal*

surface disclination points and of related surface disclination lines is given by several authors.[115,116] Some authors have argued that in the neighbourhood of a disclination line in a nematic there is a transition to the isotropic liquid phase; in other words disclination lines are pipes of isotropic liquid.[117]

Figure 30a could represent one plane (the cholesteric plane) in a cholesteric liquid crystal. At a height z above this plane the pattern would be the same

---

[114] J. Friedel and P. G. de Gennes, *Compt. rend.*, 1969, **268**, 257.
[115] P. E. Cladis and M. Kléman, *J. Phys.*, Paris, 1972, **33**, 591; R. B. Meyer, *Phil. Mag.*, 1973, **27**, 405; *Mol. Crystals Liquid Crystals*, 1972, **16**, 355.
[116] C. Williams, V. Vitek, and M. Kléman, *Solid State Comm.*, 1973, **12**, 581; M. Kléman and C. Williams, *Phil. Mag.*, 1973, **28**, 725; A. Saupe, *Mol. Crystals Liquid Crystals*, 1973, **21**, 211; T. J. Scheffer, H. Gruler, and G. Meier, *Solid State Comm.*, 1972, **11**, 253.
[117] C. M. Dafermos, *Quart. J. Mech. Appl. Math.*, 1970, **23**, S49.

but rotated through an angle $\alpha = qz$ where q is a constant. Any axis normal to the cholesteric plane with direction given by the unit vector $\hat{z}$ could be regarded as the axis of rotation: it is termed the cholesteric axis. There are three quantized symmetry operations in addition to onefold axes of rotational symmetry:[7] translation $n\pi\hat{z}/q$ (n an integer); twofold axes of rotation parallel to the cholesteric axis; twofold axes of rotation in the cholesteric plane either locally parallel to the molecules or perpendicular to them. The third operation leads to disclinations of rotation $n\pi$ rad that Friedel and Kléman[7] term τ disclinations or λ disclinations for the two sub-cases, respectively. The first two operations lead, perhaps surprisingly to the same defects, χ disclinations.[7] This can be seen by recognizing that a combination of one of the first operations with one of the second is a *non-quantized* screw symmetry operation. It follows that any dispiration with displacement $z\hat{z}$ and rotation $qz\hat{z}$ can be dissipated by plastic relaxation. In particular a disclination of rotation $n\pi\hat{z}$ rad can dissipate as a dispiration of rotation $n\pi\hat{z}$ rad and Burgers vector $n\pi\hat{z}/q$ and leave behind a dislocation of Burgers vector $-n\pi\hat{z}/q$. Thus χ disclinations in cholesteric crystals can be viewed as either dislocations with Burgers vectors $-n\pi\hat{z}/q$ or disclinations with rotations $n\pi\hat{z}$ rad. χ twist disclinations in cholesterics were first proposed by de Gennes[118] to explain Grandjean–Cano 'walls', fringe-like lines or contours seen by Grandjean[119] and Cano[120,121] in cholesterics confined between glass plates forming a low-angled wedge.

Disclinations appear not to have been observed in smectic liquid crystals.[7] If they do occur they would have rotations $n\pi$ rad about twofold axes of rotational symmetry in planes parallel to the smectic layers.[7] Dislocations in smectics have Burgers vectors normal to the smectic layers.[7]

Further information on defects in liquid crystals is to be found in numerous papers.[7,8,12,18,34,36,37,48—56,108—127]

The cuticles of beetles exhibit a structure similar to that of cholesteric crystals. Bouligand has described disclinations in them.[18,19,37,128]

---

[118] P. G. de Gennes, *Compt. rend.*, 1968, **266**, 571.
[119] F. Grandjean, *Compt. rend.*, 1921, **172**, 71; F. Grandjean, *Bull. Soc. franç. Min. Crist.*, 1919, **42**, 42.
[120] R. Cano, Thèse, Montpellier, 1966.
[121] R. Cano, *Bull. Soc. franç. Min. Crist.*, 1968, **91**, 20.
[122] Orsay Liquid Crystal Group, *Phys. Letters (A)*, 1969, **28**, 687.
[123] C. Caroli and E. Dubois-Violette, *Solid State Comm.*, 1969, **7**, 799.
[124] P. E. Cladis and M. Kléman, *Mol. Crystals Liquid Crystals*, 1972, **16**, 1.
[125] C. Williams, P. Pieranski, and P. E. Cladis, *Phys. Rev. Letters*, 1972, **29**, 90.
[126] H. Imura and K. Okano, *Phys. Letters (A)*, 1973, **42**, 405.
[127] M. Kléman, *Phil. Mag.*, 1973, **27**, 1057; *J. Phys.*, Paris, 1973, **34**, 931; I. E. Dyaloshinskii, *Soviet Phys.*, J.E.T.P., 1970, **31**, 773; C. Fan, *Phys. Letters (A)*, 1971, **34**, 335; J. L. Ericksen, in 'Liquid Crystals and Ordered Fluids', Plenum Press, 1970, p. 181; P. E. Cladis, M. Kléman, and P. Pieranski, *Compt. rend.*, 1971, **273**, 275; J. Rault, *ibid.*, 1971, **272**, 1275; *Mol. Crystals Liquid Crystals*, 1972, **16**, 143; T. J. Scheffer, *Phys. Rev. (A)*, 1972, **5**, 1327; J. Nehring, *Phys. Rev. (A)*, 1973, **7**, 1737; P. G. de Gennes, 'Introduction to the Physics of Liquid Crystals', Oxford University Press, in press.
[128] Y. Bouligand, *J. Microscopie*, 1971, **11**, 441.

4 Movement of Disclinations in Solids

Among the earliest accounts of movement of disclinations are those of Friedel[54] in nematic liquid crystals. An early treatment of movement of disclinations in solids is that of Sáenz.[129] For more comprehensive accounts of the movement of disclinations and the associated mechanical forces the reader is referred to several recent papers.[130,131] Here we shall consider the two most important cases: wedge disclinations coincident with their rotations and small twist disclination and dispiration loops encircling their rotations.

The only type of disclination known to exist in crystals other than liquid crystals is the wedge disclination coincident with its rotation (Figures 18—23): it also occurs in surface crystals (see Defects in Surface Crystals). It can move in any direction in the surface crystal by producing or removing dislocations[17,92,132] in a way reminiscent of dislocation climb involving the production or removal of vacancies or interstitials. Similar conclusions were reached independently by de Wit[27,133,134] for the same types of defects but in conventional crystals.

The mechanism[17,92,132] is illustrated for surface crystals in Figure 32. A positive wedge disclination of rotation $\pi/2$ rad is located at unit number 1 in Figure 32a. The row of units from 1 to H may be viewed as an 'extra half-row' (or half-plane, if the Figure is viewed as a section in a conventional crystal) of a nascent edge dislocation. This dislocation can glide away from the disclination along the glide line G_1. The disclination and its rotation are now located within the triangle 123 of Figure 32b. The same row from 1 to H is regarded as an 'extra half-row' of a second nascent dislocation which glides away as before along G_1 (Figure 32c and d). The net result is that the disclination has moved from unit number 1 to a neighbouring unit, number 2, *via* an intermediate position within the triangle 123. In the process it has acted as the source of two dislocations. The same displacement would have resulted if the two dislocations had glided away along G_2 or had climbed away along the row 1H or had moved away from the disclination by a mixed process of climb and glide. Again the result is the same if dislocations with the opposite Burgers vector are annihilated at the disclination. A mechanism involving climb has been described by Hasiguti.[135]

The displacement can be represented conveniently in a defect-free portion of surface crystal (Figure 33). The shaded portion represents the sector

---

[129] A. W. Sáenz, *J. Rat. Mech. Anal.*, 1953, **2**, 83.
[130] E. S. P. Das, M. J. Marcinkowski, R. W. Armstrong, and R de Wit, *Phil. Mag.*, 1973, **27**, 370.
[131] H. Günther, *Phys. Status Solidi (B)*, 1972, **49**, 551; *Exp. Tech. Phys.*, 1972, **20**, 291.
[132] W. F. Harris and L. E. Scriven, *J. Appl. Phys.*, 1971, **42**, 3309.
[133] R. de Wit, in 'Abstract Bulletin' (TMS Fall Meeting and ASM Materials Engineering Congress, Cleveland, Ohio, 1970), The Metallurgical Society of AIME, New York, and American Society for Metals, Metals Park, Ohio, 1970, p. 11.
[134] R. de Wit, ref. 1, p. 677.
[135] R. R. Hasiguti, *J. Phys. Soc. Japan*, 1973, **35**, 313.

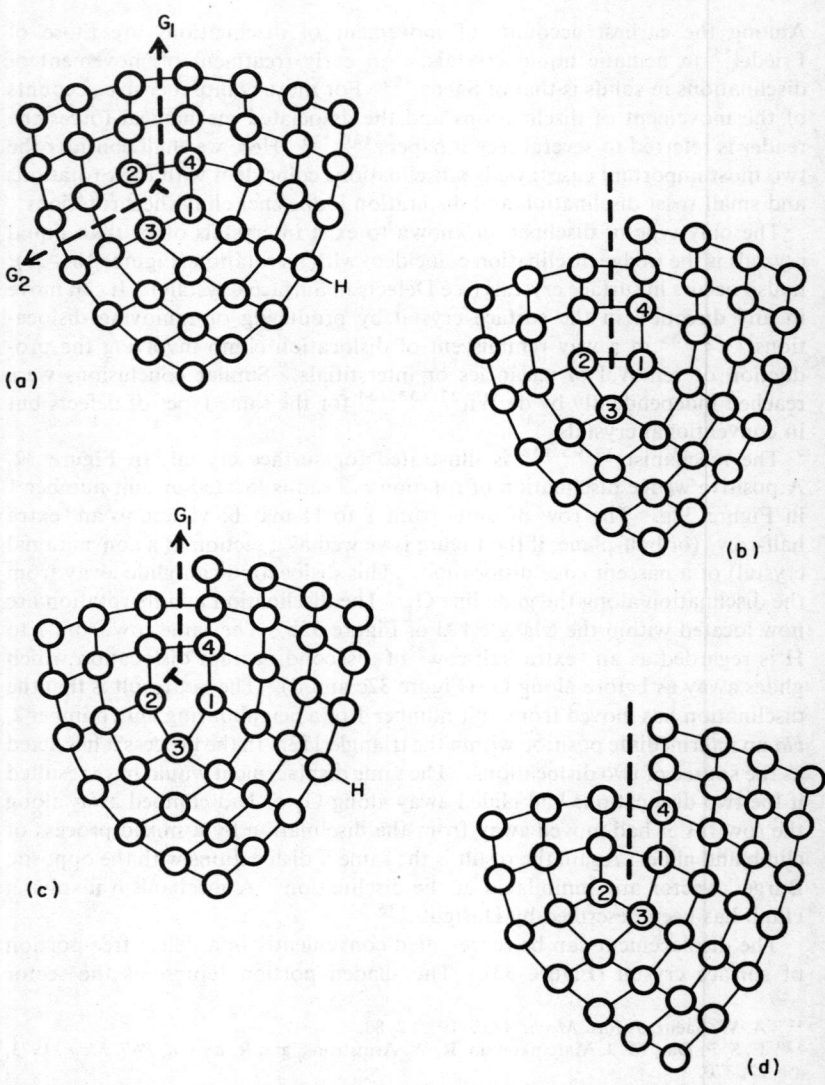

Figure 32 *A wedge disclination of rotation* $\pi/2$ *rad acting as the source of two dislocations which leave by glide along* G_1. *Located at 1 in* (a) *the disclination is displaced to the centre of the triangle 123 in* (b) *after the first dislocation leaves. A second dislocation glides away* (c) *displacing the disclination to 2 in* (d). G_2 *is an alternative glide line for the dislocations*
(Reproduced by permission from *J. Appl. Phys.*, 1971, **42**, 3309)

The Geometry of Disclinations in Crystals

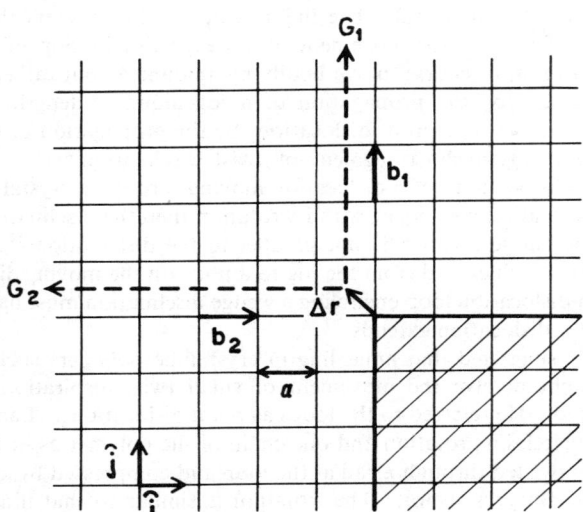

Figure 33 *Planar representation of the displacement of the disclination in Figure 32. The shaded area is the sector removed to make the wedge disclination of rotation $\pi/2$ rad; $\Delta \mathbf{r}$ is the displacement represented by (a) to (b) in Figure 32; $\mathbf{b}_1$ and $\mathbf{b}_2$ are the Burgers vectors of the dislocations corresponding to the glide lines G_1 and G_2, respectively; a is a lattice parameter*
(Reproduced by permission from *J. Appl. Phys.*, 1971, **42**, 3309)

removed in introducing the disclination situated at 1 in Figure 32a. The first step in the displacement of the disclination (Figure 32a to b) is represented by the vector $\Delta \mathbf{r}$. The full displacement from 1 to 2 could be represented by $2\Delta \mathbf{r}$. If the dislocations leave along G_1 then they have Burgers vectors given by $\mathbf{b}_1$ while if they leave along G_2 they have Burgers vectors $\mathbf{b}_2$.

In general a wedge disclination of rotation ω displaced by amount $\Delta \mathbf{r}$ to a neighbouring location will give rise to a dislocation with Burgers vector[17,92,132]

$$\mathbf{b}_1 = (\mathbf{I} - \mathbf{R}) \cdot \Delta \mathbf{r} \qquad (20)$$

or

$$\mathbf{b}_2 = \Delta \mathbf{r} \cdot (\mathbf{R} - \mathbf{I}) \qquad (21)$$

These equations can be obtained directly from equations (7) and (6), respectively, by putting $\mathbf{b} = \mathbf{0}$ and $\mathbf{a}_0 = \Delta \mathbf{r}$.

Wedge disclinations of rotation $\pi/2$ rad, the positive and negative variety, move in steps between points that are not symmetrically equivalent (Figure 32 a to b and b to d). On the other hand, wedge disclinations of rotation $\pi/3$ rad move between symmetrically equivalent points giving rise to or annihilating a dislocation at each step.[132]

Figure 32 may be viewed as a section perpendicular to a wedge disclination line in a conventional crystal. The Figure would then represent the process described by de Wit[27] for movement of a wedge disclination in a crystal. The line does not, of course, move bodily by amount $\Delta\mathbf{r}$ but rather achieves the displacement by the propagation of a jog along its length. (This is analogous to the movement of dislocations by the propagation of kinks and jogs.) The jog is a short segment of twist disclination.[27,133] From it emerges the dislocation. If a dislocation moving through a crystal intersects a wedge disclination coincident with its rotation then the disclination will be jogged if the Burgers vector is not parallel to the disclination.[27] Again a dislocation line will extend from the jog to a node on the moving dislocation. In general a dislocation loop encircling a wedge disclination must have a node from which a dislocation extends.[27]

Reneker[83] suggested that annealing in crystalline polymers such as polyethylene might involve the movement of small twist dispiration loops of rotation π rad. He referred to the loops as point dislocations. Each dispiration loop encircles its rotation and one chain of the polymer as in Figure 15. The chain is twisted through π rad at the loop and compressed by an amount given by the Burgers vector. The situation is similar to that illustrated in Figure 16b. If the Burgers vector vanishes then the loop is a twist disclination and the chain is twisted without being compressed or extended much as in Figure 1b. Nabarro has suggested that twist disclination loops are involved in muscle contraction.[136] The reporter believes that the defects are more likely to be twist dispirations.[17,69] A detailed discussion of the role of disclination loops in polymers has been given by Li and Gilman.[15]

Twist disclination and dispiration loops encircling their rotations ω move in the direction $\pm\omega$ with neither the production nor annihilation of dislocations or any other defects. The moving loop is a local region of twisting of the encircled fibre in the case of the disclination and a local region of screw displacement in the case of the dispiration. The net result of the passage along the length of a fibre is rotation of the fibre about its axis through angle ω in the first case and in the second case rotation through ω plus translation by an amount given by the Burgers vector. If the rotation ω of the dispiration loop vanishes then one has simply an edge dislocation loop. Kuo and co-workers have discussed moving disclination loops in both homogeneous and two-phase materials.[137]

Recently, Das and co-workers have described a mechanism, not unlike the Frank–Read source for dislocations, by which disclinations in a crystal can be generated.[138]

---

[136] F. R. N. Nabarro, in 'Physics of Strength and Plasticity', ed. A. S. Argon, M.I.T. Press, Cambridge, Mass., and London, 1969, p. 97.
[137] H. H. Kuo, T. Mura, and J. Dundurs, *Internat. J. Engineering Sci.*, 1973, **11**, 193.
[138] E. S. P. Das, R. de Wit, R. W. Armstrong, and M. J. Marcinkowski, *J. Appl. Phys.*, 1973, **44**, 4804.

3
Stress-induced Martensitic Transformations and Twinning in Organic Molecular Crystals

BY M. J. BEVIS AND P. S. ALLAN

1 Introduction

Stress-induced martensitic transformations and twinning can be the predominant modes of plastic deformation in crystalline polymers when they are fabricated and strained in certain well-defined ways. Although many papers on the subject of stress-induced transformations and twinning have been published, no attempt has been made to consider comprehensively the criteria which govern the operation of these deformation processes in crystalline polymers. This chapter represents an attempt to do so.

Stress-induced martensitic transformations and twinning processes involve a co-operative rearrangement of molecular chains which results in the accommodation of an applied stress. The volume of crystal which transforms martensitically undergoes a change in crystal structure as well as a characteristic change in shape. Deformation twinning results in a characteristic change in shape of the crystal but the crystal structures before and after deformation are the same.

In contrast to crystalline polymers, the operation of martensitic transformations in metals seldom occurs as a result of straining crystals of metals or alloys at room temperature. The transformations do not therefore act as deformation mechanisms, although they do play a very important part in the strengthening of these materials when introduced as a fine dispersion of second phase platelets. It will be apparent in subsequent sections of this chapter that the martensitic transformation processes in metals are in general more complex than those in polymer crystals. Twinning can be a predominant mode of plastic deformation in some metals and alloys, particularly in crystals in which only a small number of slip systems can operate.

Numerous investigations of martensitic transformations and twinning processes in both single-crystal and bulk spherulitic forms of crystalline polymers have been carried out, and there is no doubt that they can be predominant modes of plastic deformation and can consequently give rise to marked changes in crystalline texture after relatively small deformations. The early work of Frank, Keller, and O'Connor[1] led to the identification of the stress-induced martensitic transformation from the orthorhombic form to

---

[1] F. C. Frank, A. Keller, and A. O'Connor, *Phil. Mag.*, 1958, 3, 64.

the monoclinic form of polyethylene. The two structures are represented schematically in Figure 1, following the experimental results of Tanaka, Seto, and Hara.[2] In addition to presenting experimental evidence for the operation of martensitic transformations and twinning in polyethylene, Frank, Keller, and O'Connor[1] outlined the reasons why such processes should be operative in long-chain compounds, and also made many comments

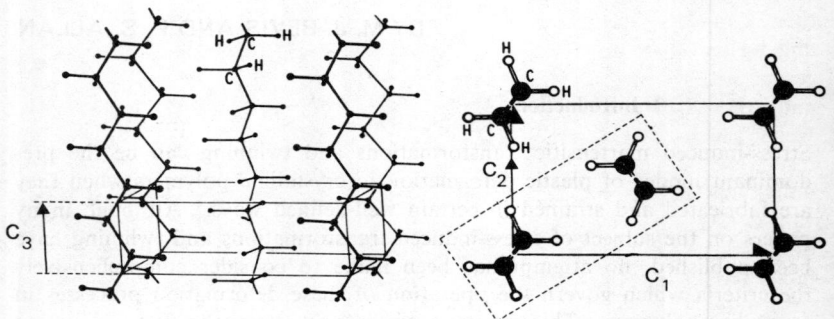

Figure 1a *Schematic diagram of the orthorhombic phase of polyethylene*
(Reproduced by permission from *Trans. Faraday Soc.*, 1939, **35**, 428)

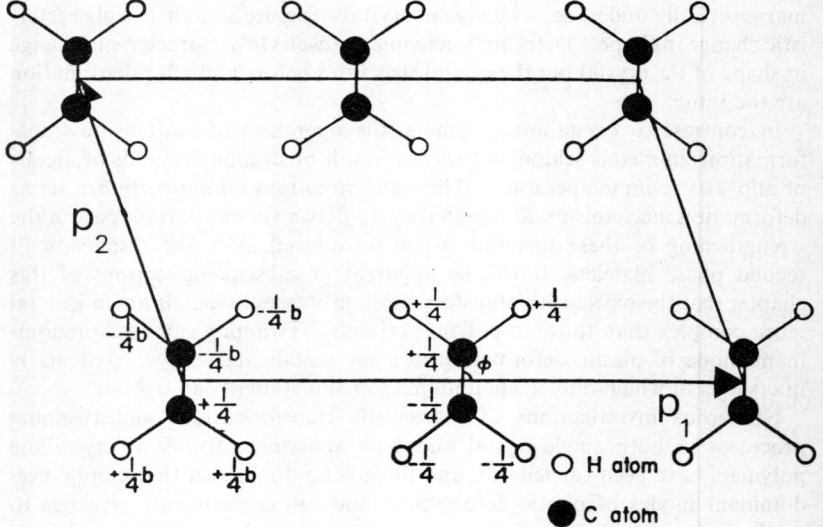

Figure 1b *Schematic diagram of the monoclinic phase of polyethylene*
(Reproduced by permission from *J. Phys. Soc. Japan*, 1962, **17**, 873)

[2] K. Tanaka, T. Seto, and T. Hara, *J. Phys. Soc. Japan*, 1962, **17**, 873.

relating to the mechanisms of nucleation and growth associated with these processes. These have in many cases been substantiated by subsequent investigations. The orientation relationships between the orthorhombic and monoclinic phases and the dependence of the transformation and twinning processes on the orientation of the strain axes relative to the crystal axes were investigated by Kiho, Peterlin, and Geil[3] using single crystals of polyethylene. The most comprehensive study to be published on twinning and martensitic transformations in bulk polyethylene is due to Seto, Hara, and Tanaka[4] in their X-ray investigation of the deformation of oriented polyethylene. Hay and Keller[5] have briefly reviewed some of the more important papers concerned with the marked changes in texture which occur at small deformations and arise as a result of twinning and transformation processes. Hay and Keller[5] included in their review some discussion of their own important results of X-ray studies on deformed nylon 66 and polyethylene which have unfortunately not been published in full. These would be particularly relevant as they represent an investigation aimed at developing a coherent picture of plastic deformation processes involving the transverse displacement of molecular chains and the role these processes play in the deformation of spherulites. One of the most comprehensive studies of the role transformation processes play in spherulite deformation is due to Yee and Stein[6] in their micro- and macro-X-ray studies of local deformation in stretched polybutene-1 spherulites. Martensitic transformations have also been reported to occur in polypropylene[7,8] and have been held accountable for changes in crystalline texture in cold-rolled polyoxymethylene.[9,10]

The papers referred to above, together with many other publications on twinning and martensitic transformations in polymer crystals, have not comprehensively considered the criteria which govern the operation of these processes. This is due in part to the absence of theoretical treatments of the crystallography of these processes and also to the aims, extent, and the type of experiments which have been made in the past. The aim of the present chapter is to summarize the evidence which exists to show that there are several criteria which certainly control the operation of martensitic transformation and twinning processes in polyethylene and which could well control the operation of these types of processes in other long-chain compounds. A lot of emphasis will be laid on the orthorhombic to monoclinic phase transformation and twinning in polyethylene which is the best documented of all the transformation and twinning processes reported to occur in polymer crystals. This chapter is mainly based on recent studies by the

---

[3] H. Kiho, A. Peterlin and P. H. Geil, *J. Appl. Phys.*, 1964, **35**, 1699.
[4] T. Seto, T. Hara and K. Tanaka, *Jap. J. Appl. Phys.*, 1968, **7**, 31.
[5] I. L. Hay and A. Keller, *J. Polymer Sci., Part C, Polymer Symposia, No. 30*, 1970, **8**, 289.
[6] R. Y. Yee and R. S. Stein, *J. Polymer Sci., Part A-2, Polymer Phys.*, 1970, **8**, 1661.
[7] D. R. Morrow and B. A. Newman, *J. Appl. Phys.*, 1968, **39**, 4944.
[8] B. A. Newman and S. Song, *J. Polymer Sci., Part A-2, Polymer Phys.*, 1971, **9**, 181.
[9] J. E. Preedy and E. J. Wheeler, *Nature Phys. Sci.*, 1972, **236**, 60.
[10] E. P. Chang, R. W. Gray, and N. G. McCrum, *J. Materials Sci.*, 1973, **8**, 397.

authors and their co-workers,[11-13] many of which are as yet unpublished in detailed form. The experimental work was designed specifically to aid the identification of the geometrical criteria governing the operation of stress-induced twinning and transformations in polyethylene and involved electron diffraction and electron microscopy studies of deformed single crystals and spherulites of polyethylene.

Crystalline polymers have some particularly important micromorphological features.[14,15] Bulk polymers exhibit a spherulitic structure which usually consists of an arrangement of interpenetrating twisted lamellae radiating from a central point. In polyethylene, for example, the [010] direction in the crystalline lamellae is always parallel to the spherulite radius. Some schematic diagrams which represent the spherulitic structure are shown in Figure 2. The lamellae within the spherulites are of the order of 100 Å in thickness, and the mechanisms of growth of the lamellae are such that their surfaces are defined by the folds in the molecular chains which give rise to the crystalline arrays within the lamellae. The orientation of the fold surface relative to the underlying molecular chain axis ([001] in the case of polyethylene) is believed to be variable. The crystal lamellae together with the folded surface layers and the volume occupied by the polymer chains which tie the lamellae together constitute a two-phase structure of amorphous and crystalline polymer chains. This is represented schematically in Figure 2 as a sandwich of amorphous and crystalline layers.

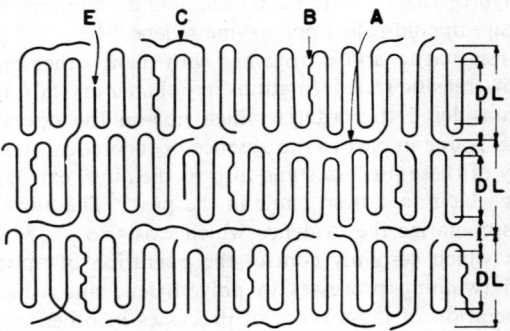

Figure 2a *Stack of densely packed parallel lamellae of the microspherulitic structure:* A, *interlamellae tie molecule*, B, *boundary layer between two mosaic blocks*, C, *chain end in 'amorphous' surface layer*, E, *linear vacancy caused by the chain end in the crystal lattice*
(Reproduced by permission from *J. Materials Sci.*, 1971, **6**, 490)

---

[11] M. J. Bevis and E. B. Crellin, *Polymer*, 1971, **12**, 666.
[12] P. S. Allan, E. B. Crellin, and M. J. Bevis, *Phil. Mag.*, 1973, **27**, 127.
[13] P. S. Allan, E. B. Crellin, and M. J. Bevis, to be published.
[14] A. Keller, *Kolloid-Z.*, 1969, **261**, 386; Reports on Progress in Physics, 1968, vol. 31 (part 2) 623.
[15] J. E. Breedon, J. F. Jackson, M. J. Marcinkowski, and M. E. Taylor, *J. Materials. Sci.*, 1973, **8**, 1071.

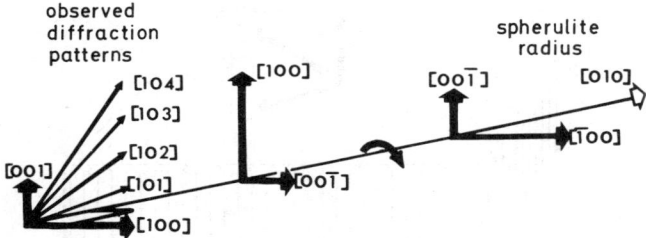

Figure 2b *Illustration of the variation in lamella orientation along the radius of a polyethylene spherulite. This coincides with the [010] direction in the polyethylene lattice. The low-index directions which are normal to the radius are illustrated*

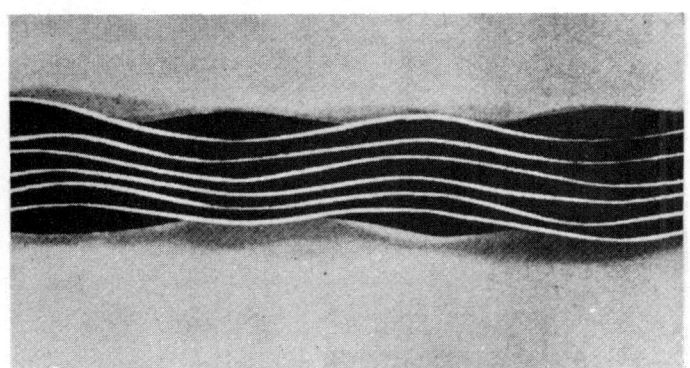

Figure 2c *Packing arrangement of lamellae in a polyethylene spherulite possessing alternating partial twists as proposed by Breedon et al. from scanning electron microscopy studies*
(Reproduced by permission from *J. Materials Sci.*, 1973, **8**, 1071)

Single crystals of many polymers can be prepared easily[16] and their morphology and thickness have been shown to be very similar to those of the lamellae referred to above. It is possible to produce, by a choice of suitable preparation conditions,[12,17] flat single crystals where the fold surface is normal to the molecular chain axis after the crystals have collapsed onto a hard substrate. The molecules fold in a characteristic way as indicated in Figure 3 to give rise to the final shape of the crystal which is consistent with the two-fold symmetry axes exhibited by the orthorhombic form of polyethylene. The geometry of the folds has been considered in detail by Burbank[18] and Keller.[14] Schematic diagrams which illustrate the orientations and positions of two possible arrangements of tight folds in the {110} sectors of a flat polyethylene crystals are shown in Figure 3.

[16] P. H. Geil, 'Polymer Single Crystal', Interscience, New York and London, 1963.
[17] D. J. Blundell, A. Keller, and A. J. Kovacs, *J. Polymer Sci., Part B, Polymer Letters*, 1966, **4**, 481.
[18] R. D. Burbank, *Bell System Tech. J.*, 1960, 1627.

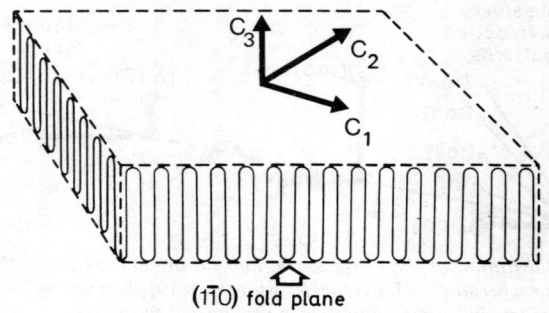

Figure 3a *Schematic diagram illustrating the relationship between fold geometry and the morphology of a diamond-shaped {110} fold-sector crystal*

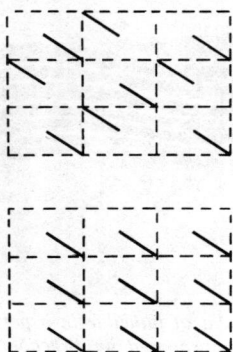

Figure 3b *Schematic diagram illustrating the orientations and positions of two possible arrangements of tight folds in the fold surface of a {110} fold sector of a polyethylene single crystal*
(Reproduced by permission from *Bell System Tech. J.*, 1960, 1627)

Figures 2 and 3 indicate two of the main differences between metal and polymer crystals which could lead to a modification of the criteria which govern the operation of plastic deformation modes in crystalline polymers.

Firstly, and with reference to Figure 2a, it is not at all clear which phase will deform preferentially. Three main types of deformation process are possible. Intra-lamella deformation describes plastic deformation processes such as slip, twinning, and martensitic transformations which occur within the lamellae, and are associated with the transverse displacement of chains. Chain-axis slip is used to describe the slip processes associated with the slip direction which is parallel to the molecular chain axis. This is widely considered to be the predominant mode of intra-lamella deformation. Inter-lamella slip describes the sliding of the lamellae relative to one another and is due to the preferential shearing of the amorphous regions between adjacent lamellae. In spherulites all three mechanisms are in competition with each other as primary modes of deformation. Substantial variation in the orientation of lamellae within spherulites relative to applied strain axes would probably result in different modes of deformation being predominant in different regions of a spherulite. There is some strong evidence to support this view.[5,6,13]

Secondly, it is feasible that intra-lamella deformation processes, which most closely resemble plastic deformation processes in metals and minerals, could be influenced by differences in fold geometry. In a single crystal of polyethylene, for example, it is clear that there will be a substantial difference in the fold plane geometry in different parts (sectors) of the crystal. This difference in geometry could give rise to a difference in the critical resolved shear stress for a particular mode of plastic deformation which could result in the operation of different plastic deformation processes in adjacent sectors of a single crystal.

In order to identify the criteria which govern the operation of stress-induced twinning and martensitic transformations, and in particular the fold sector dependence of these modes of deformation, it is essential that the complexity of the crystal morphology be minimized, and in particular that chain slip and inter-lamella deformation be eliminated. It is therefore advisable to work with flat, single-layer crystals.[12,13] To study the dependence of the modes of plastic deformation within spherulites, with respect to position in the spherulite and percentage deformation, it is essential to study the deformation of individual spherulites. The experimental techniques used in both types of experiment are reviewed in Section 4, and the experimental results and their interpretation are reviewed for single crystals and spherulites in Section 5. The discussion is restricted mainly to deformation by twinning and martensitic transformations which are the subject of this review. The basic crystallography of these processes in polyethylene and an analysis which could be used for investigating the crystallography of martensitic transformations and twinning in a large number of polymers are reviewed in Sections 3 and 2, respectively.

2 The Crystallography of Martensitic Transformations

There have been many theoretical and experimental investigations into the crystallography of martensitic transformations in metal alloys, and in particular in steels, where in some cases the occurrence of martensitic transformations is of considerable technological importance. It is of value to draw on the large amount of literature which is available on martensitic transformations in metals[19-21] and which in the case of the theories of the

---

[19] C. M. Wayman, Introduction to the Crystallography of Martensitic Transformations. (Macmillan, New York) 1964.
[20] C. M. Wayman, *Adv. Materials Res.*, 1968, **3**, 147.
[21] J. W. Christian, 'The Theory of Transformations in Metals and Alloys', Pergamon Press, Oxford, 1965.

crystallography of martensitic transformations apply in principle to all crystalline materials.

The total shape change F associated with a martensitic transformation should on a macroscopic scale, by definition, be an invariant plane strain. The transformation process does not proceed by diffusion. Figure 4 is a

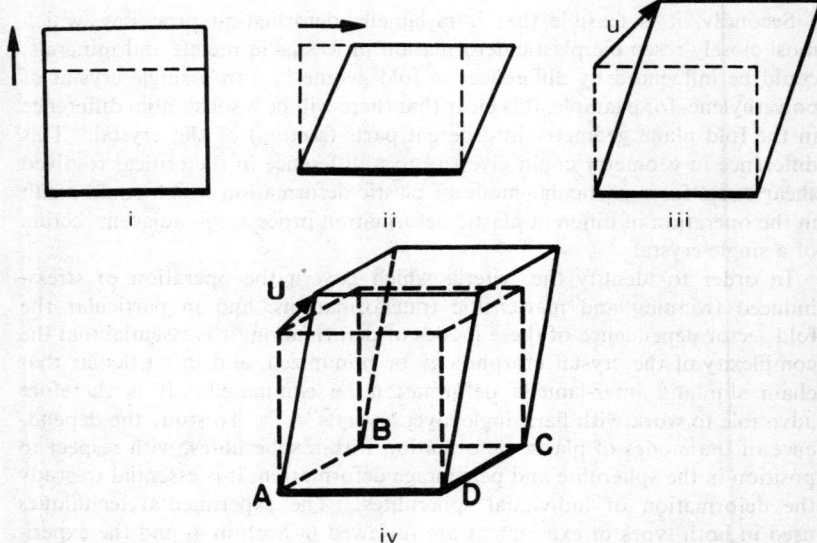

Figure 4 *Schematic diagrams of the shape change associated with a martensitic transformation* (i) *simple extension,* (ii) *simple shear,* (iii) *and* (iv) *general invariant plane strain. The original shapes are represented by broken lines and the invariant planes are represented by bold lines*

schematic diagram of the shape change associated with a martensitic transformation and this can in general be described as the product of a pure strain P and a rigid body rotation R, so that

$$F = RP$$

The pure strain P describes the way in which the product structure can be obtained from the parent by expansion or contraction along three mutually orthogonal axes such that, when referred to these axes, the components of P are

$$\begin{pmatrix} e_1 & 0 & 0 \\ 0 & e_2 & 0 \\ 0 & 0 & e_3 \end{pmatrix}$$

This, of course, takes into account any volume change associated with the transformation. The pure strain together with a rigid body rotation must be equivalent to an invariant plane strain. However, it may be shown from straightforward geometrical considerations that the following restrictions on the values of the principal strains e_1, e_2, and e_3 must be imposed[19-21] if a plane is to be left undistorted. One of the principal strains must be zero, one less than zero and one greater than zero, e.g. $e_3 = 0$, $e_2 > 0$, $e_1 < 0$. The principal strains for a particular transformation are determined by the lattice parameters of the parent and product structures and a choice of correspondence matrix which gives a description of the relationships between vectors in both phases before and after transformation. A fuller description of these points is given by Wayman.[19] Having identified a correspondence which is realistic in that it involves relatively small values of e_i and which relates the two structures closely, it is possible, in principle, to determine the invariant plane and shape shear direction associated with the transformation.

The observed macroscopic shape change associated with martensitic transformations in ferrous alloys approximates closely to an invariant plane strain. However, the principal strain matrix calculated for a choice of Bain correspondence, which is generally agreed to be the correspondence which is associated with these transformations, has principal strains such that none of them is equal to zero. In fact two are equal and are greater than zero, and the third is less than zero. The restrictions on these principal strains for the total shape change to be an invariant plane strain are not satisfied. However, this discrepancy can be accounted for if, over and above the pure strain, a lattice invariant deformation S is introduced. This deformation causes no change in structure but does result in a shape change which, together with P and a rigid body rotation R, is such that the product RPS approximates on a macroscopic scale to an invariant plane strain. A way in which it is possible to cause a change in the shape of a crystal without changing the structure is illustrated in Figure 5. There are excellent examples of the agreement between experimental results and the results of theoretical treatments based on this model.[22] A lattice invariant deformation system is one which is likely

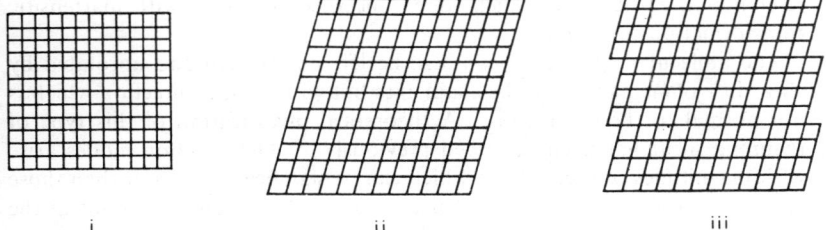

Figure 5 (i) *Initial crystal*, (ii) *after lattice deformation*, (iii) *lattice deformation followed by slip shear as the lattice invariant deformation* (After Bilby and Christian, Inst. Met. Monograph No. 18, 1955, p. 121.)

[22] D. S. Liberman, in 'The Mechanism of Phase Transformations in Crystalline Solids', Institute of Metals (London) Monograph and Report Series No. 33, 1969, p. 167–175.

to operate as a deformation mode in the parent or product, and may be an operative slip system or a deformation twinning mode. However, the internal structure associated with transformations in metals can be very complex, and there are still some outstanding problems associated with a full description of the transformation process in some cases. In principle it is possible for a multiple-shear lattice invariant deformation to operate, and theories which take this into account have been developed.[23]

Martensitic Transformations in Crystalline Polymers.—Polymer crystals differ from metal crystals in one important way with respect to martensitic transformations.[1] The crystal axis which is parallel to the strong covalently bonded polymer chains is unlikely to be distorted by the transformation process, and must therefore be parallel to the invariant plane of the transformation. In a martensitic transformation in polymer crystals each molecular chain will in general have to undergo a rotation about its own axis. The transverse displacement of chains satisfies automatically the condition that $e_3 = 0$. Thus, the requirements for an invariant plane strain in this case are only that $e_2 > 0$ and $e_1 < 0$. These conditions are satisfied for many choices of correspondence matrices which are consistent with the transverse displacement of chains, and the conditions for an invariant plane strain are satisfied without the introduction of a lattice invariant deformation in the form of twinning or slip. Indeed, the incorporation of a lattice invariant deformation would result on application of the martensite crystallography theories[23] for these choices of correspondence matrices in a lattice invariant shear of magnitude zero. Martensitic transformations in polymers therefore differ from the majority of transformations that occur in metals and alloys in that in many cases the incorporation of a lattice invariant deformation is not required. The crystallography of martensitic transformations in polymer crystals is therefore much more straightforward than that in metals and alloys. The treatment presented below is comprehensive and, with the aid of examples presented thereafter, could be applied to polymer systems without reference to earlier texts which were concerned primarily with martensitic transformations in metals.

The molecular chain axes can also be left undistorted and unrotated by shear in a plane parallel to the chain axes but with a shear direction which is not normal to the chain axes. A theoretical investigation of this type of twinning shear has been described previously.[11] Deformation modes arise with magnitudes of shear strain which can be significantly greater than those associated with the transverse displacement of chain axes, but result in the same orientation relationships. Under suitable conditions of the orientation of an applied uniaxial stress such modes could be generated even though the magnitude of the shear strain is greater than those modes associated with the transverse displacement of chains.

[23] A. F. Acton, M. J. Bevis, A. G. Crocker, and N. D. H. Ross, *Proc. Roy. Soc.*, 1970, **A320**, 101.

The constraints described above restrict martensitic transformations to polymer crystal structures with space lattices which exhibit two-fold symmetry axes. These axes must be parallel to the molecular chain axes in both structures and must be undistorted and unrotated by the transformation process. For any choice of correspondence which is consistent with the transverse displacement of chains, a martensitic transformation is possible in theory between two structures with lattices exhibiting symmetry axes 2, 4, or 6, provided that a symmetry axis is parallel to the chain axis in both structures, and that the repeat distance along one axis is an integral multiple of the repeat distance along the other. In this situation the crystallography is essentially two dimensional, as demonstrated and discussed in detail[11] for the case of the orthorhombic to monoclinic transformation in polyethylene. The theoretical treatment[11,23] used for that transformation applies for all of the class of transformations described above, and the method of application is described below. In situations where one of the phases associated with a transformation has a triclinic space lattice, there must exist a special relationship between the two lattices in order that the transformation can proceed without the incorporation of a lattice invariant deformation. This is the situation for the γ- to α-phase transformation in polypropylene which has been considered previously.[13]

The Crystallography of 'Two-Dimensional' Martensitic Phase Transformations.—The most general two-dimensional transformation which can occur is one in which the parent lattice C and product lattice P are defined by basis vectors $\mathbf{c}_1$, $\mathbf{c}_2$ and $\mathbf{p}_1$, $\mathbf{p}_2$ with included angles α and β, respectively. The polymer chain axes in the parent and product phases are taken to be parallel to $\mathbf{c}_3$ and $\mathbf{p}_3$, respectively, which are parallel to each other and also normal to $\mathbf{c}_1$, $\mathbf{c}_2$, $\mathbf{p}_1$, and $\mathbf{p}_2$. The metrics of these bases are given by

$$c_{ij} = \mathbf{c}_i \cdot \mathbf{c}_j = \begin{pmatrix} (c_1)^2 & c_1 c_2 \cos \alpha \\ c_1 c_2 \cos \alpha & (c_2)^2 \end{pmatrix};$$

$$c^{ij} = \mathbf{c}^i \cdot \mathbf{c}^j = \begin{pmatrix} (c_1 \sin \alpha)^{-2} & -(c_1 c_2 \sin^2 \alpha)^{-1} \cos \alpha \\ -(c_1 c_2 \sin^2 \alpha)^{-1} \cos \alpha & (c_2 \sin \alpha)^{-2} \end{pmatrix}$$

$$p_{ij} = \mathbf{p}_i \cdot \mathbf{p}_j = \begin{pmatrix} (p_1)^2 & p_1 p_2 \cos \beta \\ p_1 p_2 \cos \beta & (p_2)^2 \end{pmatrix};$$

$$p^{ij} = \mathbf{p}^i \cdot \mathbf{p}^j = \begin{pmatrix} (p_1 \sin \beta)^{-2}, & -(p_1 p_2 \sin \beta)^{-1} \cos \beta \\ -(p_1 p_2 \sin \beta)^{-1} \cos \beta & (p_2 \sin \beta)^{-2} \end{pmatrix}$$

where $\mathbf{c}^1$, $\mathbf{c}^2$ and $\mathbf{p}^1$, $\mathbf{p}^2$ are the reciprocal bases to $\mathbf{c}_1$, $\mathbf{c}_2$ and $\mathbf{p}_1$, $\mathbf{p}_2$ respectively, and $\mathbf{c}^i \cdot \mathbf{c}_j = \delta^i_j$ and $\mathbf{p}^i \cdot \mathbf{p}_j = \delta^i_j$, respectively, where δ^i_j is the Kronecker delta and is equal to 1 when $i = j$ and 0 when $i \neq j$.

The correspondence matrix describes the relationship between vectors before and after the transformation and essentially defines a pair of transformation modes. The vector $[0, 1]_C$ after transformation becomes $[u^1, u^2]_P$

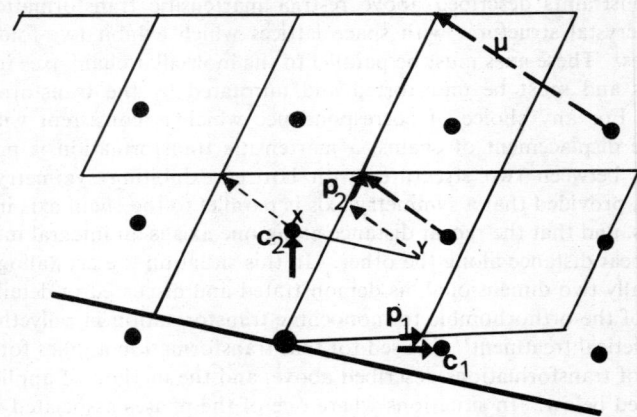

Figure 6 *Schematic diagram of an invariant plane strain which transforms the unit cell defined by c_1, c_2 into the cell defined by p_1, p_2: The original lattice is represented by the circles and the product lattice is defined by the points of intersection of the two arrays of parallel lines. The trace of the invariant plane is represented by a bold line and the shape deformation direction is indicated by μ. The correspondence in this example is $\begin{pmatrix} 1 & 1 \\ 0 & 1 \end{pmatrix}$ so that $[1,0]_c$ $[0,1]_c$ transform to $[1,0]_p$ and $[\bar{1},1]_p$ respectively. The distance which a point moves as a result of the transformation process is proportional to the perpendicular distance of that point from the invariant plane. Points at X and Y for example move the same distance and in a direction parallel to μ.*

and the vector $[1, 0]_C$ after transformation becomes $[u^1{}_2, u^2{}_2]_P$. A list of correspondence matrices which are most likely to give rise to a 'two-dimensional' transformation mode with a shape deformation of small magnitude has been given previously.[11] For a particular choice of correspondence the magnitude of the shape deformation, f, the shape deformation direction, $u^1 c_1 + u^2 c_2$, and the invariant plane normal, $h_1 c^1 + h_2 c^2$, are given by the solutions of the following equations. The metrics, the elements of a correspondence matrix, the invariant plane and the shape deformation direction are illustrated in Figure 6 for a two-dimensional transformation not involving a lattice invariant transformation.

$$(h_2/h_1)^2 c_{11} - (p_{11} u_1^1 u_1^1 + 2 p_{12} u_1^1 u_1^2 + p_{22} u_1^2 u_1^2)$$
$$- 2(h_2/h_1) c_{21} - (p_{11} u_1^1 u_2^1 + p_{12} u_1^1 u_2^2 + p_{12} u_2^1 u_1^2 + p_{22} u_1^2 u_2^2)$$
$$+ c_{22} - (p_{11} u_2^1 u_2^1 + 2 p_{12} u_2^1 u_2^2 + p_{22} u_2^2 u_2^2) = 0 \qquad (1)$$

$$(u^2/u^1)^2 c^{11} - (p^{11} u_2^2 u_2^2 - 2 p^{12} u_2^2 u_2^1 + p^{22} u_2^1 u_2^1)$$
$$- 2(u^2/u^1) c^{12} - (p^{11} u_2^2 u_1^2 + p^{12} u_2^2 u_1^1 + p^{21} u_1^2 u_2^2 - p^{22} u_2^1 u_1^1)$$
$$+ c^{22} - (p^{11} u_1^2 u_1^2 - 2 p^{12} u_1^2 u_1^1 + p^{22} u_1^1 u_1^1) = 0 \qquad (2)$$

$$f^2 = c^{11}(p_{11}u_1^1u_1^1 + 2p_{12}u_1^1u_1^2 + p_{22}u_1^2u_1^2) + 2c^{12}p_{11}u_1^1u_2^1$$
$$+ p_{12}(u_1^1u_2^2 + u_2^1u_1^2) + p_{22}u_1^2u_2^2$$
$$+ c^{22}(p_{11}u_2^1u_2^1 + 2p_{12}u_2^1u_2^2 + p_{22}u_2^2u_2^2) \qquad (3)$$
$$- 2(p_1p_2 \sin \beta)/(c_1c_2 \sin \alpha).$$

The correct pairing of the solutions $[u^1, u^2]$ and (h_1, h_2) to the quadratic equations satisfies the following relationship:

$$F = 1 + f(\bar{u}^1\bar{h}_1 + \bar{u}^2\bar{h}_2) \qquad (4)$$

where F is the volume change associated with the transformation, and $\bar{u}^1, \bar{u}^2$ and $\bar{h}_1, \bar{h}_2$ are the components of the unit vectors parallel to the vectors $[u^1, u^2]$ and (h_1, h_2), and are given by $[\bar{u}^1, \bar{u}^2] = [u^1, u^2]/(c_{11}u^1u^1 + 2c_{12}u^1u^2 + c_{22}u^2u^2)^{\frac{1}{2}}$ and $(\bar{h}_1, \bar{h}_2) = (h_1, h_2)/(c^{11}h_1h_1 + 2c^{12}h_1h_2 + c^{22}h_2h_2)^{\frac{1}{2}}$.

For every correspondence listed in Table 1, and in general, it is possible to

Table 1 *The crystallographic parameters associated with the twenty most probable martensitic transformation modes in polyethylene. The fraction $1/m$ is the fraction of the Bravais lattice points of the parent phase which are sheared to their correct positions in the product phase*

N	U	Ū	m	First invariant plane	Second invariant plane	First shape deformation direction	Second shape deformation direction	f
T1	$\begin{pmatrix} 1 & 0 \\ 0 & 1 \end{pmatrix}$		2	1,3.78,0	−4.67,1,0	−3.11,1,0	1,5.33,0	0.201
T2	$\begin{pmatrix} 1 & 0 \\ 1 & 1 \end{pmatrix}$	$\begin{pmatrix} -1 & 1 \\ -1 & 0 \end{pmatrix}$	1	1,2.35,0	15.54,−1,0	2.57,−1,0	1,12.50,0	0.318
T3	$\begin{pmatrix} 1 & 0 \\ 0 & 1 \end{pmatrix}$		1	1,−2.32,0	16,75,1,0	2.52,1,0	1,−13.48,0	0.343
T4	$\begin{pmatrix} 0 & 1 \\ -1 & 1 \end{pmatrix}$		2	2.54,−1,0	1.44,1,0	1,2.45,0	1,−1.39,0	0.437
T5	$\begin{pmatrix} -1 & 0 \\ 0 & 1 \end{pmatrix}$		2	1,1.26,0	−1.29,1,0	−1.23,1,0	1,1.31,0	0.764
T6	$\begin{pmatrix} 1 & 0 \\ 2 & 1 \end{pmatrix}$		1	1,1.26,0	47.35,−1,0	1.29,−1,0	1,38.09,0	0.960
T7	$\begin{pmatrix} 1 & 0 \\ -1 & 1 \end{pmatrix}$	$\begin{pmatrix} 1 & -1 \\ 0 & 1 \end{pmatrix}$	1	48.57,1,0	1,−1.24,0	1,−39.08,0	1.26,1,0	0.984
T8	$\begin{pmatrix} 0 & 1 \\ -1 & 1 \end{pmatrix}$	$\begin{pmatrix} 1 & 1 \\ 0 & 1 \end{pmatrix}$	1	−1.21,1,0	1,1.52,0	1,1.23,0	−1.49,1.0	1.024
T9	$\begin{pmatrix} 1 & 0 \\ -1 & 1 \end{pmatrix}$		2	1.17,1,0	8.20,−1,0	1,−1.15,0	1,7.88,0	1.029
T10	$\begin{pmatrix} 0 & 1 \\ 1 & 0 \end{pmatrix}$		1	1,22,1,0	−1,1.57,0	−1,1.24,0	1.54,1,0	1.042

generate sixteen correspondence matrices by changing the signs and interchanging the rows and columns of the matrix. When the 'two-dimensional' parent and product lattices are of higher symmetry than 2, the number of matrices which are not crystallographically equivalent is reduced. The most straightforward way of establishing whether or not the condition for an invariant plane is satisfied, that is $e_2 > 0$, $e_1 < 0$ or $e_2 < 0$, $e_1 > 0$, is to determine the values of (h_2/h_1). In cases where it is impossible to transform the parent into the product without the incorporation of a lattice invariant deformation, the solutions to the quadratic equations listed above are imaginary.

The Crystallography of 'Two-dimensional' Deformation Twinning.—The crystallography of deformation twinning is more straightforward than the crystallography of martensitic transformations and is therefore better understood.[24,25] The structures of the parent and product in deformation twinning are identical so that $p_1 = c_1$ and $p_2 = c_2$. Equations (1)—(4) apply equally well to twinning, and indeed twinning and martensitic transformation processes have been treated together in a comprehensive treatment of the crystallography of co-operative shear-like processes in crystals.[23] Equation (4) does, however, become redundant because, in twinning, the shape deformation direction lies in the twin plane so that $\bar{u}^1 \bar{h}_1 + \bar{u}^2 \bar{h}_2 = 0$. The solutions can be correctly paired from an analysis of the angles between the two twinning directions and the normals to the two twinning planes resulting from the application of the analysis, and usually referred to as η_1, η_2 and K_1, K_2, respectively. The angles between K_1 and η_2, K_1 and K_2, and η_1 and K_2 must be acute and the angle between η_1 and η_2 must be obtuse.[24,25] A detailed application of the crystallography of twinning to all crystal systems has been published.[25]

Comparison of Theory and Experiment.—In the application of the analyses summarized above, correspondences which relate directly all or a large fraction of the lattice points of the Bravais lattices of the two phases are those considered most likely to describe operative twinning and transformation processes. The procedure for applying the analysis is to determine from equations (1)—(4) the habit planes, the shape deformation directions, and the magnitudes of the shape deformation for the correspondences which are most likely to describe operative martensitic transformation and twinning processes. The modes that will be expected to operate are those associated with a small magnitude of shear.[21,26]

The confirmation of the crystallography of a martensitic transformation through a comparison of theory and experiment can only be considered to be complete when all of the crystallographic variables referred to are shown

---

[24] M. J. Bevis and A. G. Crocker, *Proc. Roy. Soc.*, 1968, **A304**, 123.
[25] M. J. Bevis and A. G. Crocker, *Proc. Roy. Soc.*, 1969, **A313**, 509.
[26] B. A. Bilby and A. G. Crocker, *Proc. Roy. Soc.*, 1965, **A228**, 240.

to be in agreement. The invariant plane should be identified by trace analysis, the shape deformation direction and the magnitude of the shape deformation should be identified from tilts produced at free surfaces by the transformation,[20,21] and finally the orientation relationship between the two phases must be identified by diffraction techniques. Clearly, the first two measurements are not practical in relation to transformations or twins occurring within polymer spherulites and indeed would be difficult to measure in polymer single crystals. The first, however, can now be contemplated with the advent of scanning transmission electron microscopes which allow polymers which are very sensitive to electron beams to be examined more readily at high magnification and at high resolutions compared with conventional electron microscopes.[27] The only really practical measurement that could be made to date is then the determination of the orientation relationships between the two phases by electron microscopy or X-ray diffraction techniques. Fortunately, the simple nature of martensitic transformations in polymer crystals is such that these measurements do allow an effective although not a complete comparison to be made between theoretical and experimental studies. Further confirmation of the crystallography of the transformation and twinning proceesss has been gained from a consideration of the resolved shear strain and fold sector dependence of the deformation modes.[12,13]

3 Application of the Martensite and Twinning Crystallography Theories

The treatment of the crystallography of twinning and phase transformations presented above only takes into account the space lattices of the parent and product structures. In geometrical treatments of the type described which were developed initially to predict twinning and transformation modes in metals it is assumed that the motif unit shears as a whole and the structure is restored by atomic shuffles where necessary. A motif unit for the polyethylene structure is enclosed in broken lines in Figure 1. It was realized[12] for the case of polyethylene that when all pairs of (CH_2—CH_2) molecular units are replaced by points then a considerable simplification of the crystallography of both twinning and martensitic transformations results. It was therefore convenient to use pseudo-space lattice bases of the type indicated in Figure 7 by $\bar{c}_1$, $\bar{c}_2$ and $\bar{p}_1$ and $\bar{p}_2$, respectively. The transformation of correspondences from the crystal lattice basis (U) to the pseudo-lattice basis ($\bar{U}$) may be shown to be:

$$\begin{pmatrix} \bar{u}_{11} & \bar{u}_{12} & 0 \\ \bar{u}_{21} & \bar{u}_{22} & 0 \\ 0 & 0 & 1 \end{pmatrix} = \begin{pmatrix} 1 & 1 & 0 \\ -1 & 1 & 0 \\ 0 & 0 & 1 \end{pmatrix} \begin{pmatrix} u_{11} & u_{12} & 0 \\ u_{21} & u_{22} & 0 \\ 0 & 0 & 1 \end{pmatrix} \begin{pmatrix} \tfrac{1}{2} & -\tfrac{1}{2} & 0 \\ \tfrac{1}{2} & \tfrac{1}{2} & 0 \\ 0 & 0 & 1 \end{pmatrix} \quad (5)$$

[27] M. J. Bevis, M. N. Thompson, and P. S. Allan, unpublished results.

and

$$\begin{pmatrix} \bar{u}_{11} & \bar{u}_{12} & 0 \\ \bar{u}_{21} & \bar{u}_{22} & 0 \\ 0 & 0 & 1 \end{pmatrix} = \begin{pmatrix} 2 & 0 & 0 \\ 0 & 1 & 0 \\ 0 & 0 & 1 \end{pmatrix} \begin{pmatrix} u_{11} & u_{12} & 0 \\ u_{21} & u_{22} & 0 \\ 0 & 0 & 1 \end{pmatrix} \begin{pmatrix} \tfrac{1}{2} & -\tfrac{1}{2} & 0 \\ \tfrac{1}{2} & \tfrac{1}{2} & 0 \\ 0 & 0 & 1 \end{pmatrix} \quad (6)$$

for twinning and phase transformations respectively.

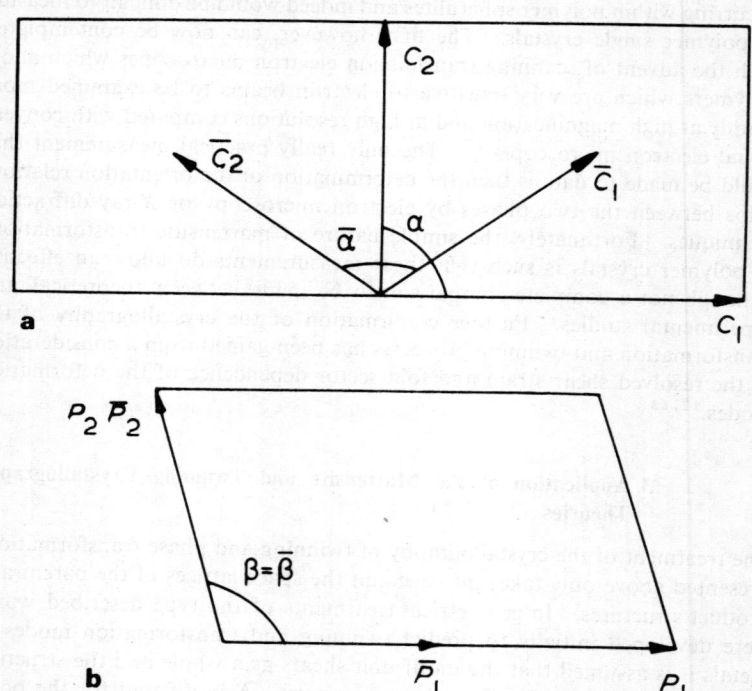

Figure 7 (a) *Parent lattice basis vectors for both twinning and phase transformations. In the case of twinning the product lattice basis vectors are the same as for the parent lattice but differ in orientation:* $c_1 = 7.41$ Å; $c_2 = 4.94$ Å; $\alpha = 90°$; $\bar{c}_1 = \bar{c}_2 = 4.45$ Å; $\bar{\alpha} = 112.6°$. (b) *The product lattice basis vectors for phase transformations:* $p_1 = 2\bar{p}_1 = 8.09$ Å; $p_2 = \bar{p}_2 = 4.79$ Å; $\beta = \bar{\beta} = 107.9°$

The martensite crystallography theory described has been used to determine the crystallographic variables associated with twinning and transformation modes with magnitudes of shape deformation less than or approximately equal to 1.5 and 1, respectively. The correspondences which result in a one to one correspondence between Bravais lattice points of parent and product structures or molecular chains in parent and product structures were the only ones which were considered. The correspondence matrices and the crystallographic variables for the predicted transformation and twinning modes are given in Tables 1 and 2. The deformation mode is referred to by the form given in the first column in the respective tables. Subscripts 1 and 2 are used to distinguish between the mode listed and the reciprocal modes (second invariant plane and direction associated with a correspondence) which is an equally feasible deformation mode. The correspondence matrices in columns 2 and 3 in Tables 1 and 2 are related by equations (5) and (6),

Table 2 The crystallographic parameters of the twelve most probable twinning modes in orthorhombic polyethylene. The magnitude of shear associated with the twinning modes is given by s

N	U	$\bar{U}$	m	K_1	K_2	η_1	η_2	s
D1		$\begin{pmatrix} -1 & 1 \\ 0 & 1 \end{pmatrix}$	2	310	$1\bar{1}0$	$1\bar{3}0$	110	0.25
D2	$\begin{pmatrix} 1 & 0 \\ 1 & 1 \end{pmatrix}$		1	120	100	$2\bar{1}0$	010	0.67
D3	$\begin{pmatrix} 0 & 1 \\ 1 & 0 \end{pmatrix}$	$\begin{pmatrix} -1 & 0 \\ 0 & 1 \end{pmatrix}$	1	$1\bar{1}0$	110	110	$\bar{1}10$	0.83
D4		$\begin{pmatrix} 1 & 1 \\ 0 & 1 \end{pmatrix}$	2	$\bar{1}10$	1,1.26,0	110	−1.26,1,0	1.08
D5	$\begin{pmatrix} -1 & -1 \\ 1 & 0 \end{pmatrix}$		1	1,4.2,0	3.37,1,0	−4.2,1,0	−1,3.37,0	1.3
D6	$\begin{pmatrix} 1 & 0 \\ 1 & 1 \end{pmatrix}$	$\begin{pmatrix} 2 & -1 \\ 1 & 0 \end{pmatrix}$	1	110	100	$1\bar{1}0$	010	1.33

respectively. A space in column 2 indicates that all molecular units are sheared to their correct positions but that the lattice is not restored by the shear. A space in column 3 indicates that the shear mode takes all Bravais lattice points to their correct positions but that all molecular chains are not restored to their correct positions. A matrix in both columns 2 and 3 indicates that all Bravais lattice points and molecular chains are restored to their correct positions by the shear process, although as in all examples the molecular chains will have to undergo a change in orientation in order to comply with the observed orientation relationships.

The $(001)_c$ plane of shear plots of the transformation shear modes $T1_{1,2}$–$T10_{1,2}$ and the deformation twinning modes $D1_{1,2}$–$D6_{1,2}$ are shown in Figures 8 and 9, respectively. They show the positions of molecular chain axes before and after deformation, and indicate that orientation relationships other than reflection in the twin plane or rotation of 180° about the twin direction are possible.[11,24]

The predicted orientation relationships between the orthorhombic and monoclinic phases for the transformation modes $T1_{1,2}$–$T10_{1,2}$ are summarized in Table 3.

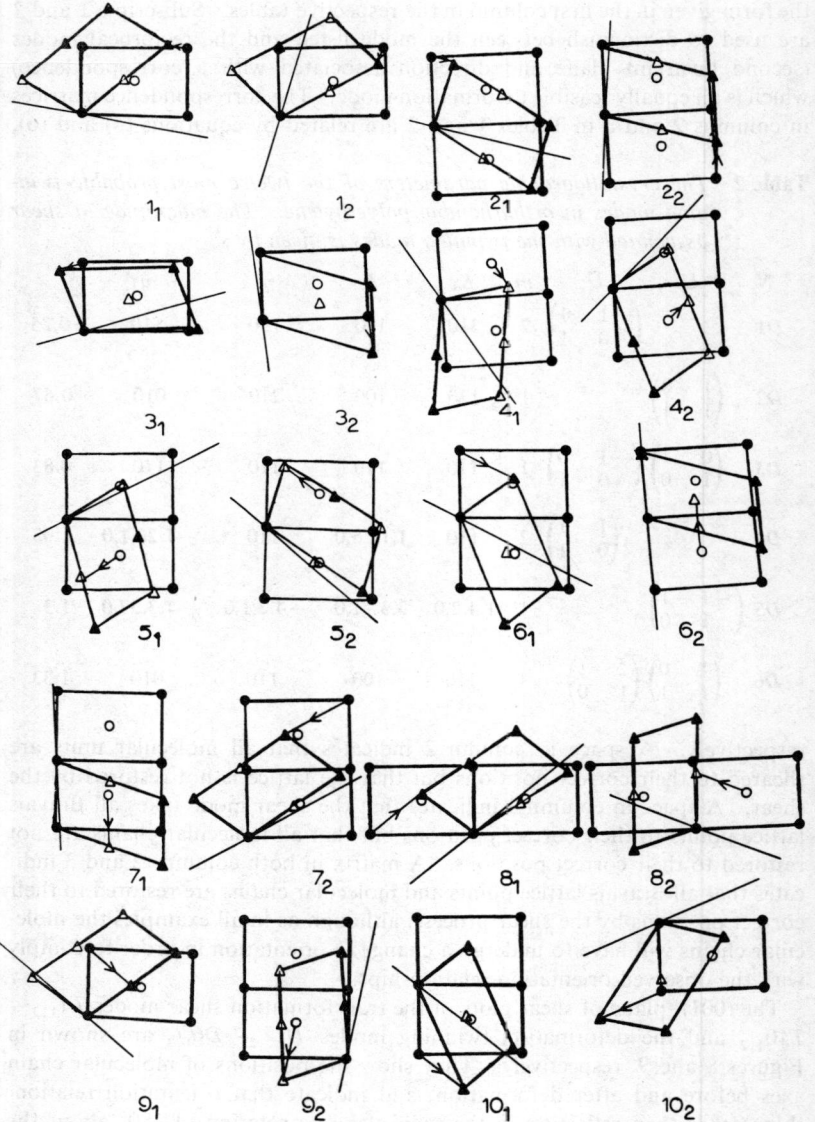

Figure 8 *Plane of shear plots for the transformation modes* $T1_{1,2}$—$T10_{1,2}$. *The* $(CH_2$—$CH_2)$ *molecular units of the parent and product structures are represented by circles and triangles, and* ● *and* ○ *shear to* ▲ *and* △, *respectively*

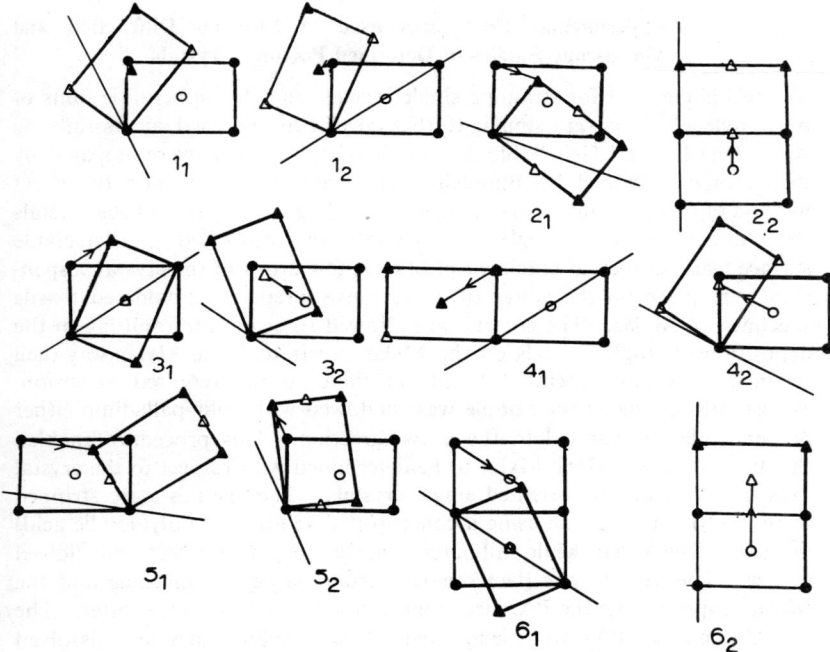

Figure 9 *Plane of shear plots for the twinning modes* $D1_{1,2}$—$D6_{1,2}$. *The notation used is the same as that in Figure 8*

Table 3 *The orientation relationships associated with the twenty transformation modes described in Table 1*

Mode	$T1_1$	$T1_2$	$T2_1$	$T2_2$	$T3_1$	$T3_2$	$T4_1$
$010_c \wedge 010_p$	122.1°	133.56°	17.80°	179.79°	160.83°	179.8°	80.24°
Mode	$T4_2$	$T5_1$	$T5_2$	$T6_1$	$T6_2$	$T7_1$	$T7_2$
$010_c \wedge 010_p$	55.66°	56.67°	15.03°	51.03°	0.0°	0.0°	127.82°
Mode	$T8_1$	$T8_2$	$T9_1$	$T9_2$	$T10_1$	$T10_2$	
$010_c \wedge 010_p$	128.97°	74.91°	123.96°	69.68°	49.71°	104.56°	

The results presented above apply only to polyethylene and represent the deformation modes which are most likely to operate on the basis of the criteria which have been identified as governing the operation of shear-like processes in metals. The analysis, subject to the restrictions summarized in Section 2, could be applied to the Bravais lattices of product and parent structures of any crystalline polymer. In cases where the motif unit consists of two or more molecular chains it may not, however, be possible to adopt the procedure used for polyethylene of considering the pseudo-space lattices obtained by replacing the molecular chains by points.

4 Experimental Procedures used in Electron Diffraction and Microscopy Studies of Deformed Polymer Crystals

The technique used for straining single crystals and thin spherulitic films of polyethylene[12,13] is very similar to that developed and used successfully by Kiho, Peterlin, and Geil.[3] Single crystals of polyethylene were prepared by the method described by Blundell, Keller, and Kovacs,[17] and the exact method of preparation[12] was chosen so that a large proportion of the crystals were diamond-shaped single-layer crystals and possessed no detectable surface features such as cracks or pleats. A few drops of the crystal suspension were place on the gauge of a previously prepared shouldered tensile specimen of Mylar. The solvent was allowed to evaporate resulting in the deposition of single crystals on the Mylar substrate. The Mylar was then strained in a hand-operated tensile machine to the required extension. Whilst still in tension the sample was shadowed with gold–palladium either perpendicular or parallel to the draw direction. This procedure enabled the draw direction of the Mylar to be determined with respect to the crystal axes of the randomly oriented single crystals. The crystals were stripped from the Mylar substrate using a concentrated solution of poly(acrylic acid) in water. The Mylar, while still in tension, was coated with PAA and allowed to dry. The sample was then removed from the tensile machine and the Mylar stripped from the PAA leaving the coated crystals on the latter. The crystal side of the PAA was then coated with carbon and the replica dissolved in a dish of distilled water which left the carbon film with the crystals floating on the surface. Parts of the carbon film were put onto grids for examination in an electron microscope.

Thin spherulitic films of high-density polyethylene were solvent cast onto Mylar, and deformed and prepared for electron microscopy using a procedure very similar to that used for the preparation of single crystals. Prior to straining, the solvent-cast films were first heated to 150 °C for approximately thirty minutes and then recrystallized at 120 °C for thirty minutes and cooled to room temperature in air. Films which exhibited a large sheaf morphology in an optical microscope were selected and subsequently strained while in contact with the Mylar substrate to 15, 20, 33, 40, and 60 per cent.

Single crystals and thin spherulitic films were examined in a conventional electron microscope operated at 100 kV and at low intensities to minimize electron beam damage to which many polymers are susceptible.[28,29] The most definite information that could be gained from the electron microscope investigations was that obtained from analysis of selected area electron diffraction patterns obtained from adjacent sectors in individual single crystals and adjacent regions in individual spherulites.

The electron diffraction patterns obtained from deformed single crystals

---

[28] D. T. Grubb and G. W. Groves, *Phil. Mag.*, 1971, **24**, 815.
[29] D. T. Grubb, A. Keller, and G. W. Groves, *J. Materials Sci.*, 1972, **7**, 131.

of polyethylene when the incident electron beam is approximately parallel to the molecular chain axis may be readily indexed. It is not quite so straightforward to index selected area diffraction patterns obtained from films exhibiting a full spherulitic structure. The lamellae orientation varies along the radius of a spherulite in the way indicated in Figure 3d. All of the low-index diffraction patterns associated with the zone axes indicated are to be expected, and the observed diffraction patterns support this and are always of the type shown in schematic form in Figure 10. Figure 10 represents the

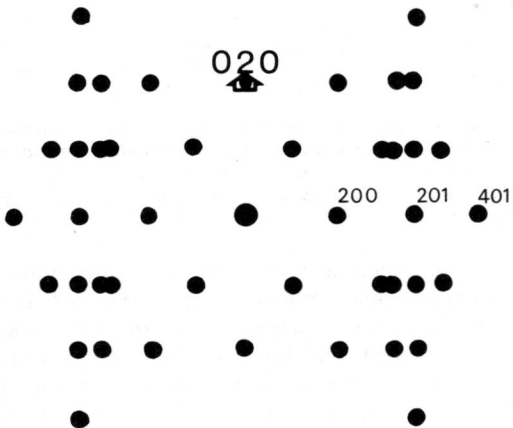

Figure 10 Schematic diagram of the selected-area electron diffraction pattern that would be obtained from a polyethylene spherulite exhibiting the lamellae orientations shown in Figure 2b. The 020 reflection would be much more intense than all other reflections obtained from an undeformed spherulite, and the line joining the centre of the pattern to the 020 reflection would be parallel to the spherulite radius

superposition of all six patterns indicated in Figure 3b, all of which have a common 020 reflection because the (020) plane is always normal to the spherulite radius and therefore always approximately parallel to the incident electron beam. By monitoring the change in the form of electron diffraction pattern shown in Figure 10 it is possible to study over most of the range of lamella orientations, the changes in texture which occur with percentage deformation and position within a spherulite.

5 Experimental Results and Discussion

The analysis of electron diffraction patterns obtained from individual sectors in more than one hundred polyethylene single crystals are summarized with respect to the orientation relationships between orthorhombic and monoclinic phases in Figure 11. The predicted orientation relationships for martensitic transformations $T1_1$ and $T2_1$ are represented by broken lines

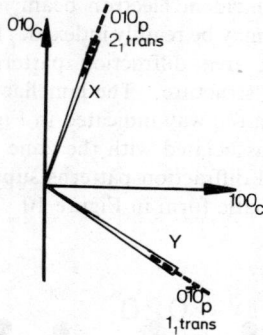

Figure 11 *The predicted and experimentally determined orientation relationships associated with stress-induced martensitic transformations in polyethylene single crystals. The predicted orientations are represented by broken lines and the angular range of experimentally determined orientation relationships are represented by triangles X and Y*

and the spread of the experimentally determined orientation relationships are represented by the triangles X and Y. It is clear that there is excellent agreement between theory and experiment. In addition to the two orientation relationships associated with the martensitic transformations, a twin orientation relationship of reflection in the {110} orthorhombic planes was also observed.[12,13] For certain limited angular ranges of the straining axis relative to the crystallographic axes two modes of deformation, $T2_1$ transformation and {110} twinning, were observed. An indexed electron diffraction which was representative of this situation and which serves as an example of the type of diffraction pattern obtained from polyethylene single crystals is shown in Figure 12.

The knowledge of the orientation relationships between parent and product structures and therefore the crystallography of the operative transformation modes allowed the resolved shear strains associated with the different deformation systems to be calculated. The variation in type of operative deformation mode with straining axis orientation is summarized in Figure 13a. The resolved shear strains associated with the different deformation systems were proportional to the values given in Figure 13b for the full range of values of θ, the angle between the straining axis and the crystal axis c_1. It is clear from a comparison of Figures 13a and 13b that a resolved shear strain criterion governs the operation of the twinning and transformation modes which have been observed to operate in single crystals deformed while in contact with a substrate. Although the experimental results described above are consistent with a resolved shear strain criterion they could also be consistent with a resolved shear stress criterion. So far it has not been possible to identify uniquely which of these criteria governs the operation of the deformation modes in the anisotropic polyethylene crystals.

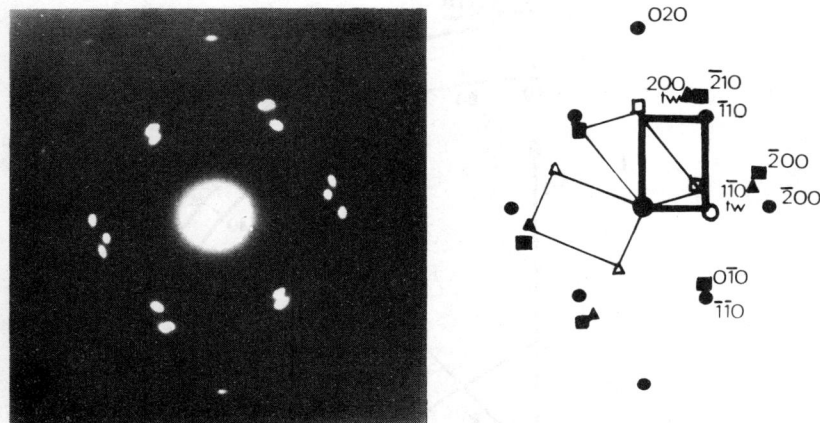

Figure 12 *Indexed electron diffraction pattern illustrating the type of pattern obtained from deformed polyethylene single crystals exhibiting the original orthorhombic phase, a twinned orthorhombic phase and the 2_1 martensite phase transformation. The open symbols represent missing reflections*

The results reported above do not take into account differences in the behaviour of adjacent sectors within a single crystal. They only show that, for a particular value of θ, there is a substantial volume within the single crystal which has undergone deformation by twinning or transformation.

The excellent agreement between the observed and predicted orientation relationships associated with the transformations $T1_1$ and $T2_1$, and the apparent resolved shear strain dependence of these deformation modes strongly indicates that the orthorhombic–monoclinic transformations are martensitic in nature. Schematic diagrams which represent the shear processes associated with the martensitic transformation and twinning shear processes are shown in Figures 14, 15, and 16.

The transformation modes $T1_1$ and $T2_1$, which are the only transformation modes to be observed in the electron diffraction experiments, possess some unique crystallographic features which are not present in any of the other transformation modes listed in Table 1. Firstly, there exists a low-index plane in the orthorhombic phase which is closely parallel to a low-index plane in the monoclinic phase. The parallel planes are $(130)_c//(410)_P$ and $(130)_c//(\bar{2}30)_P$ in the transformations $T1_1$ and $T2_1$, respectively. Secondly, the plane identified in the orthorhombic phase has the same form in both transformations. Thirdly, the low-index planes which are parallel to each other also have, to a close approximation, the same interplanar spacings. The interplanar spacings are $d_{(130)} = 1.607$ Å, $d_{(410)} = 1.605$ Å and $d_{(\bar{2}30)} = 1.59$ Å. In addition, the invariant planes associated with the transformations and indicated in Figures 14 and 15 by broken lines may be represented

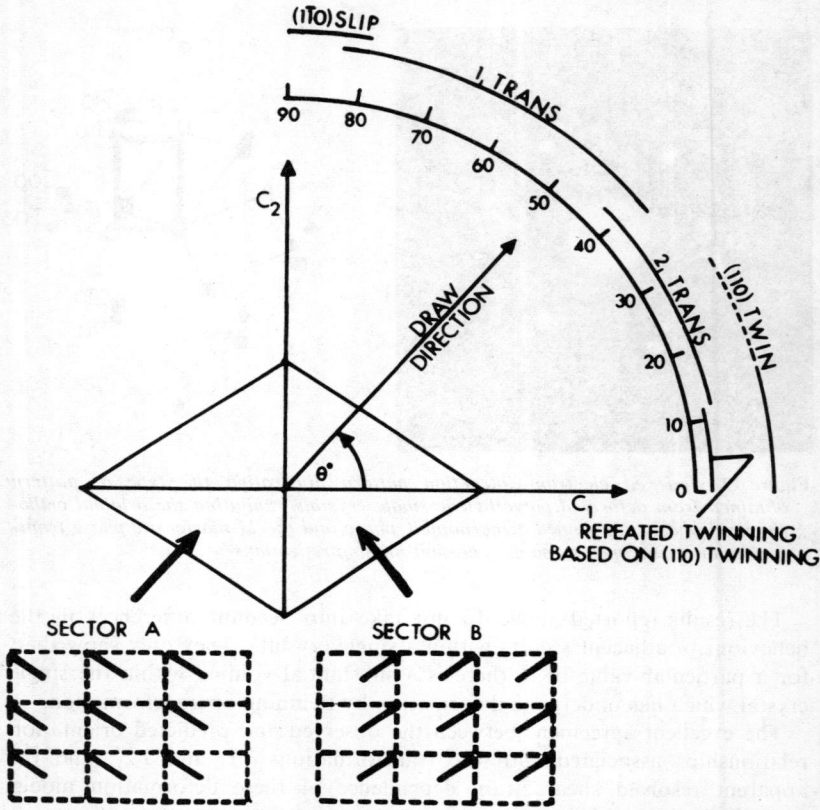

Figure 13a *The experimentally determined dependence of the operative slip, twinning, and transformation modes with θ. The notation used for identifying the different fold geometries within a single crystal relative to the straining axis (draw direction) is also illustrated*

to a good approximation by a stepped (130) interface.[12] The parallelism of low-index planes and their similar interplanar spacings mean that slip–dislocation–transformation boundary interactions which result in the formation of transformation dislocations are feasible as in the case of slip–dislocation–twin-boundary interactions.[30–32]

Although the transformation modes $T1_2$ and $T2_2$ are defined by the same correspondence and have the same magnitude of shape deformation as $T1_1$ and $T2_1$, respectively, they are not observed to operate for values of θ where the resolved shear strain on these systems is high.

[30] A. W. Sleeswyk and C. A. Verbraak, *Acta Metall.*, 1961, **9**, 917.
[31] K. Ishii and H. Kiho, *J. Phys. Soc. Japan*, 1963, **18**, 1122.
[32] M. Bevis and D. I. Tomsett, *Phil. Mag.*, 1969, **19**, 129, 533.

Martensitic Transformations and Twinning in Organic Crystals

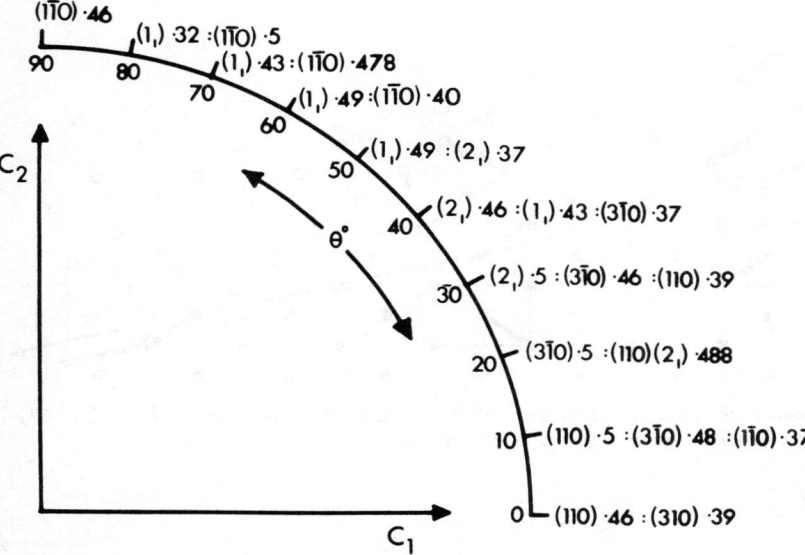

Figure 13b *The resolved shear strain dependence of the operative slip, twinning, and martensitic transformation modes*

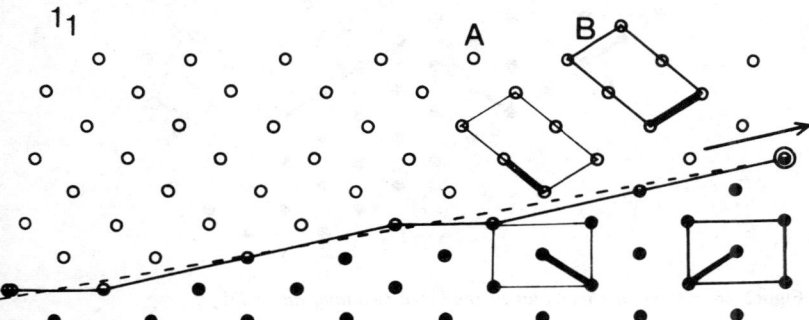

Figure 14 *Plane of shear plot of the martensitic transformation mode 1_1. The open and filled circles represent the position at which the molecular chains intercept the (001) plane of both phases. The (001) planes are represented by the plane of the diagram. The difference in orientations of the molecular chains in each of the parent and product structures has been omitted for simplicity. The irrational invariant plane associated with the transformation is represented by a broken line. The unit cells in the diagram show the changes in the fold geometry which occur in in sectors A and B following the notation introduced by Figure 13a*

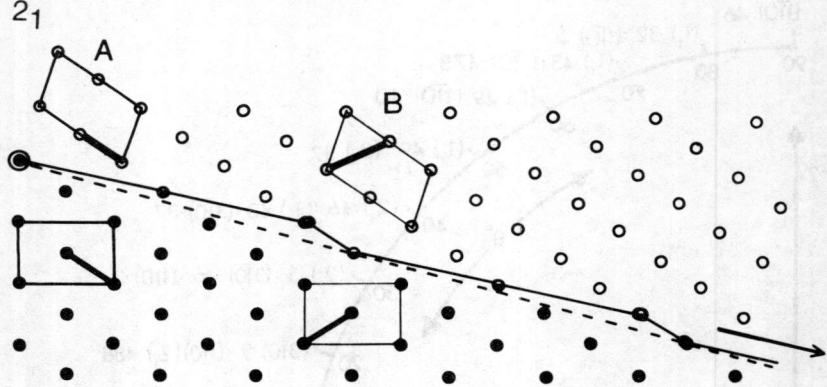

Figure 15 *Plane of shear plot of the martensitic transformation mode* 2_1. *The notation is the same as that used in Figure 14*

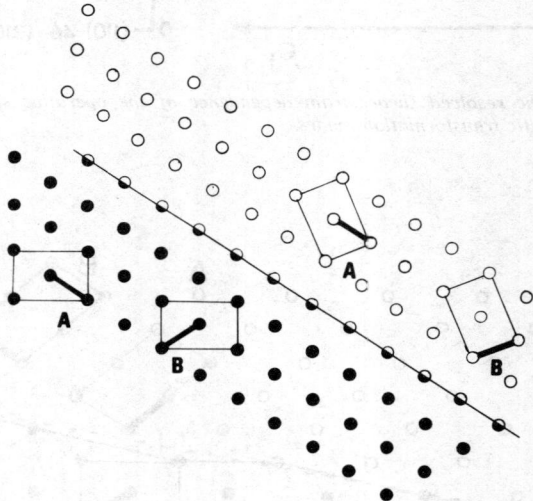

Figure 16 *Plane of shear plot of the* {110} *twinning mode* $D1_2$

It is therefore reasonable to conclude that the unique crystallographic features described above may be responsible for the preferred operation of transformation modes $T1_1$ and $T2_1$. These features could result in the preferential generation of transformation dislocations and therefore the easy growth of the transformed regions under a suitable applied strain. In addition, or alternatively, it may be that the crystallography could be consistent with the preferential nucleation of martensites in the vicinity of stacking faults associated with partial dislocations formed during the growth or deformation of the polyethylene single crystals, following the proposals made by Frank, Keller, and O'Connor[1] and Holland.[33] Presumably, the stacking faults in both cases would be on the {130} planes.

There is no simple reason why {310} twinning was not observed in the single-crystal experiments even though the twinning system is defined by the same correspondence as the {110} twins. In terms of a resolved shear strain criterion the {310} twinning is in direct competition with the $T2_1$ transformation. There is strong evidence to suggest as indicated below that the critical resolved shear strain for twinning is much less than that for martensitic transformations, so that the exclusion of {310} twinning must be due to some other factor which is unknown at present.

The criteria which govern the operation of twins in metals, that is the operative twins, usually have a small magnitude of shear strain and simple shuffles, clearly apply in the operation of transformation modes $T1_1$ and $T2_1$ and {110} twinning. The magnitudes of the shape deformations associated with the three modes are small, the smallest possible which are consistent with the requirement that no inhomogeneous shuffles[24-26] of the molecular chains be associated with the deformation process, if, as shown in Figures 8, 9, 14—16, the difference in orientation of the molecular chains is not taken into account.

A marked difference in the deformation behaviour of adjacent sectors within a single crystal was observed[13] when electron diffraction patterns were recorded individually from adjacent sectors. The differences in deformation behaviour for different values of θ for 15 percent deformation are summarized in Figure 17, from which it is clear that fold geometry has a

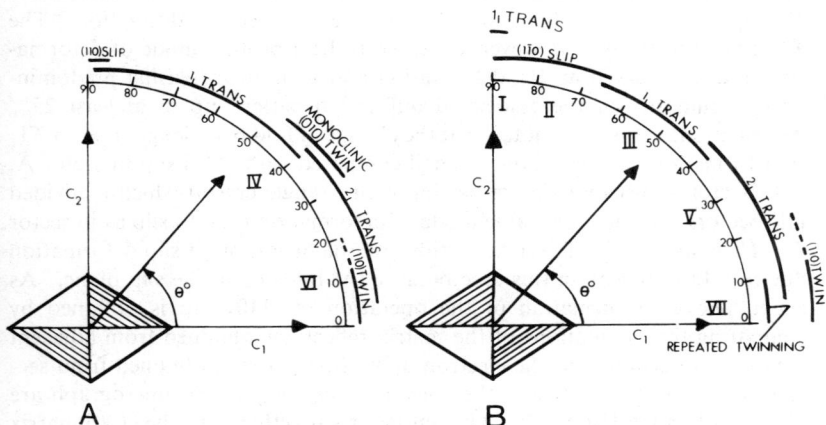

Figure 17 *The fold-sector dependence of the operative deformation modes*

[33] V. F. Holland, *J. Appl. Phys.*, 1964, **35**, 3235.

considerable effect on the operative deformation processes within individual crystalline lamellae. A prior knowledge of the crystallography of the deformation processes allows the fold-sector dependence of the deformation processes to be explained in a relatively straightforward manner.

The fold-sector dependence of the operative deformation modes in {110}-sector polyethylene single crystals is determined primarily by the orientation of the fold plane relative to the straining axis. This may be shown by detailed reference to just three of the seven zones[13] I—VII of different types of deformation behaviour referred to in Figure 17. In zone II ($\theta = 67$—$85°$) the deformation process in sector B is predominantly due to $(1\bar{1}0)$ slip as determined from the increasing difference in the relative orientations of the orthorhombic [001] patterns obtained from the two {110} fold sectors. In sector A the deformation is predominantly due to the $T1_1$ martensitic transformation.

The expected mode of deformation in sector A on the basis of a maximum resolved shear strain would be $(1\bar{1}0)$ slip as indicated in Figure 13b. However, the operation of this mode of deformation would result in an alteration of the fold geometry from $(1\bar{1}0)$ to (010) following the passage of a unit dislocation. This process does not change the fold geometry to a greater extent than that due to the operative {110} deformation twinning modes and is therefore likely to operate. Subsequent propagation of a dislocation on the same plane would require a greater extension of the fold bend to $(3\bar{1}0)$ and would increase more with the propagation of further dislocations. The fold bend lengths are represented by bold lines in Figures 14, 15, and 16. Substantial deformation by slip on the $(1\bar{1}0)$ planes in sector A is therefore unlikely. In terms of a critical resolved shear strain criterion, unit slip on $(1\bar{1}0)$ would be expected to operate first, followed by $T1_1$ martensitic transformation. The operation of unit slip cannot be identified by electron diffraction. The $T1_1$ transformation is, however, observed to be a profuse mode of deformation and is observed at 7% deformation and with an increasing predominance in intensity of the associated diffraction pattern up to at least 25% deformation. It is to be noted that the change in fold bend length due to $T1_1$ transformation is comparatively small compared with $(1\bar{1}0)$ slip in sector A.

In sector B, because we are dealing with a single crystal which is divided into sectors, the most probable mode of deformation is $(1\bar{1}0)$ slip as in sector A. There are no significant restrictions on the operation of slip deformation due to fold geometry in this case because the folds lie in the slip plane. As stated above, confirmation of the operation of $(1\bar{1}0)$ slip is obtained by comparing the orientations of the matrix reflections obtained from adjacent sectors. An example of the electron diffraction patterns obtained from sectors A and B in zone II and the corresponding bright field micrograph are shown in Figures 18a and b. The tensile axis together with the [100] matrix direction for each sector are indicated on the bright field micrograph.

As θ decreases in value from zone II to zone III there is a decrease in the resolved shear strain on the $(1\bar{1}0)$ slip system. The resolved shear strain on the (110) slip system is very low. The maximum resolved shear strain

Martensitic Transformations and Twinning in Organic Crystals

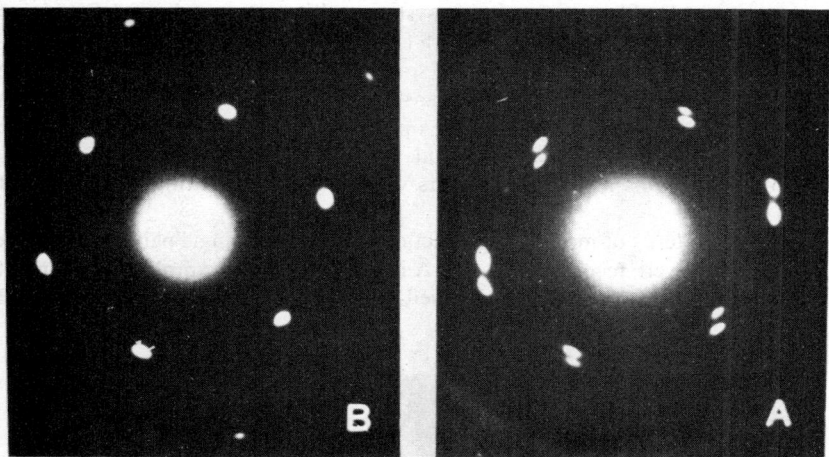

Figure 18a *Electron diffraction patterns obtained from sectors B and A, respectively, in region* II

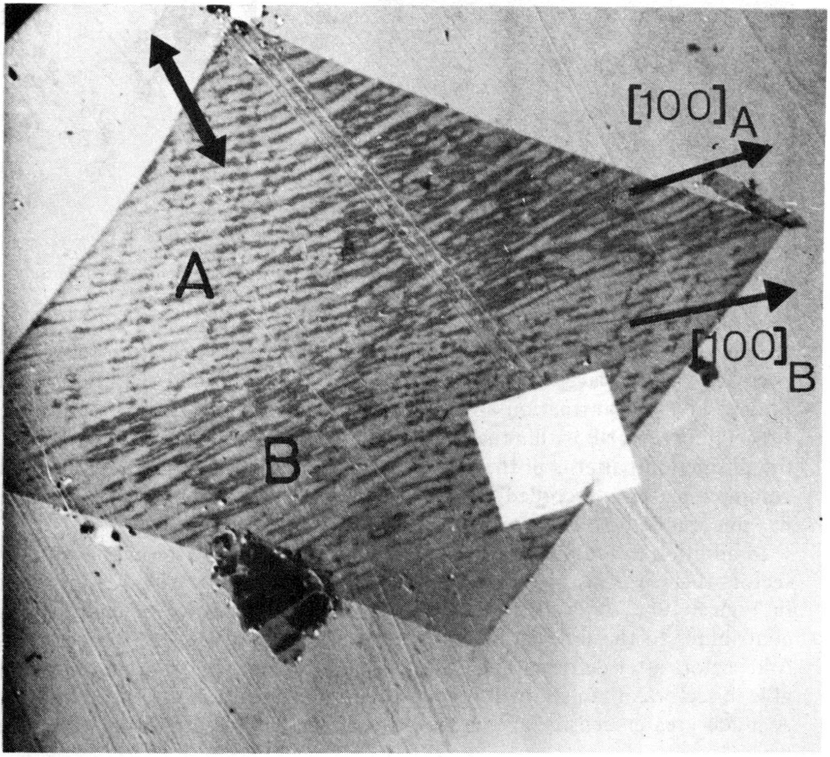

Figure 18b *Transmission electron micrograph of the single crystal from which the electron diffraction patterns in Figure* 18a *were obtained*

occurs on the $T1_1$ transformation system and it is this mode of deformation which is observed to operate in both {110} sectors. In the absence of alternative slip systems it appears that the small difference in the bend lengths of the folds, which are formed within transformed regions in the adjacent sectors of the crystal, does not restrict the operation of the transformation mode to just a single sector of a single crystal. No significant separation of the matrix orientations in adjacent sectors was observed, indicating the absence of {110} slip.

Two patterns of monoclinic reflections as well as a single pattern of matrix reflections were recorded in sector A in zone IV. An example of the type of electron diffraction pattern observed is shown in Figure 19a. In sector B

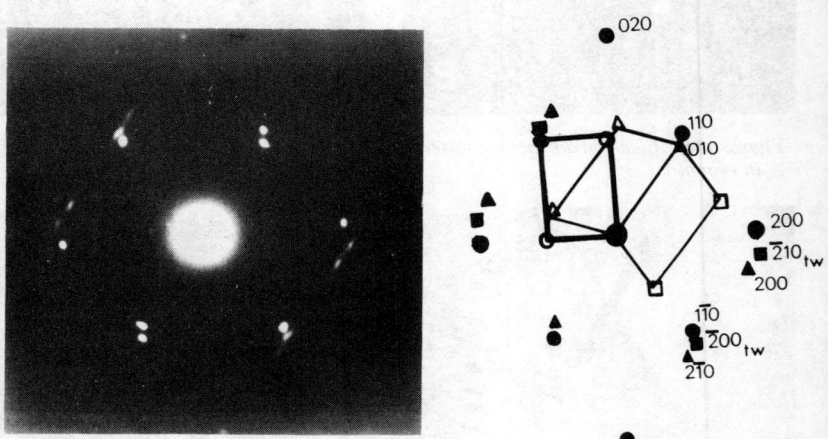

Figure 19a *An indexed electron diffraction pattern of the type observed in sector A in zone* IV

the pattern containing the additional $T2_1$ twinned transformation reflections occurred only at large strains. In sector A the additional reflections were present at 7% deformation and were always much more intense than those observed in sector B within the same single crystal. Careful measurement of the diffraction patterns of the type shown in Figure 19a shows that the most common patterns recorded represent a region of 2_1 transformation which has undergone further deformation by (010) twinning.

In addition to a difference in the recorded diffraction patterns for adjacent sectors there was also a marked difference in their morphology as indicated in Figure 19b. It is likely that a fold sector dependence of chain tilting contributes to the differences observed in the morphology of the opposing fold sectors after deformation. Slip parallel to [001] would be more favourable in sector A than sector B for the direction of strain shown in Figure 19b. A much greater density of fine cracks is observed in sector B in which the

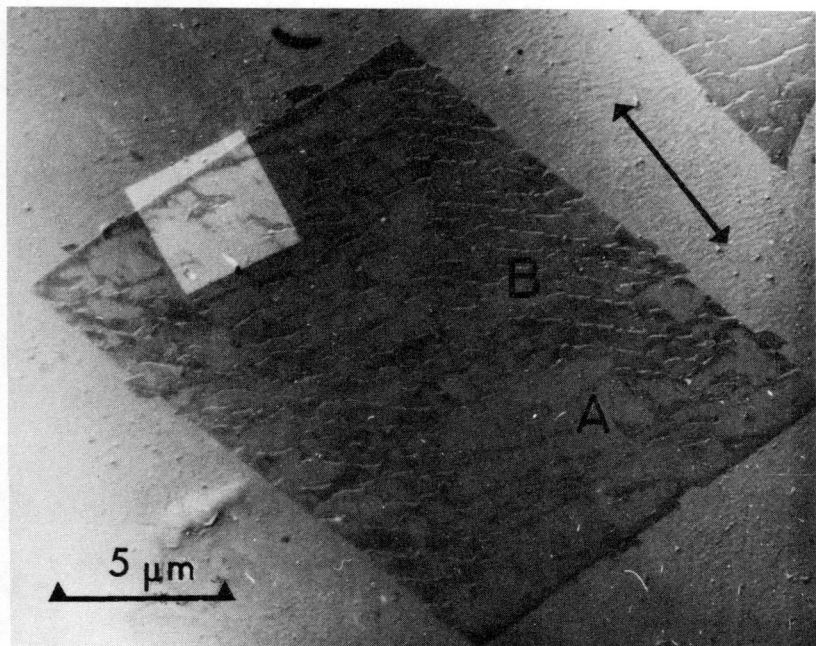

Figure 19b *Transmission electron micrograph of the polyethylene single crystal from which the diffraction pattern shown in Figure 19a was obtained*

additional deformation by twinning does not occur. If the crystal does not crack in sector B because of strong adhesion of the crystal to the Mylar substrate the additional deformation by twinning is observed, but nevertheless was always predominant in sector A.

The observed difference in behaviour of sectors A and B can be explained in terms of the fold plane geometries within the $T2_1$ transformed regions in the adjacent sectors. In the transformed regions in sector A there is no change in fold geometry as a result of (010) twinning. The large extension of the fold bend lengths which would occur as a result of (010) twinning within the $T2_1$ transformed regions in sector B is illustrated in Figure 20. The preference for monoclinic twinning in sector B is attributed to this marked difference in fold geometry. Less strain can be accommodated by the operation of deformation mechanisms in sector B than in sector A thus accounting for the preferred cracking in sector B.

The very brief consideration given above to the effect of fold geometry on deformation processes in polyethylene single crystals demonstrates that there are two important factors which control the operation of the observed deformation processes. Firstly, the deformation process must not result in large extensions of the folded parts of the molecular chains. This is clearly

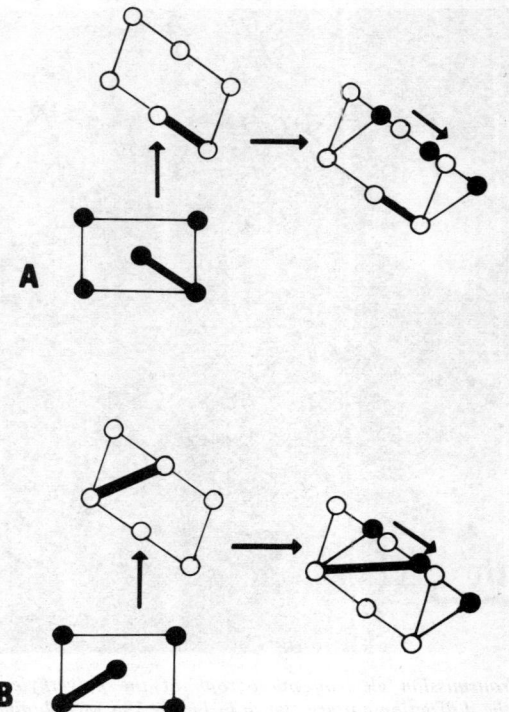

Figure 20 *Schematic diagrams showing the changes in fold geometry which occur in fold-sectors* A *and* B *after the operation of the multiple transformation and twinning processes reported to occur in zone* IV

apparent from the analysis of both zones II and IV and applies to deformation by twinning, martensitic transformation, and slip. The experimental results presented can be simply explained in terms of tight folds and therefore support the case for the existence of this form of chain folding. A comparison of the fold bend lengths, projected onto the (001) plane, which result from (010) twinning in the $T2_1$ martensites in sectors A and B in zone IV, indicates that the maximum extension possible must be between 16 and 60%. These are the increases in length of the projected fold bend lengths after $T2_1$ transformation in sector B and $T2_1$ transformation followed by (010) twinning in sector B, respectively. The change in the projected fold bend length for the transformation plus twinning mechanisms in sector A results in a 9.8% decrease. The second factor which controls the operation of deformation processes is apparent from a consideration of zones II and III. The changes in fold bend length associated with $T1_1$ transformation and $(1\bar{1}0)$ slip in sector B in zone II are not significantly different. The critical resolved shear strain for slip must therefore be less than that for the martensitic transformation. In fact there will in many situations be competition between the recorded deformation processes based on the difference between the applied shear strain and the magnitude of the critical resolved shear strain required to operate a particular deformation process. In sector B in zone III the operative deformation mode is the 1_1 transformation because the resolved shear strain on this system is greater than the critical resolved shear strain for transformation, but less than the critical resolved shear strain for extensive slip. The behaviour in zone V may be considered as a further example; here the $T2_1$ transformation is predominant in both sectors because the resolved shear strain on this system is high compared with that of the possible slip systems which leave the fold plane invariant.

The deformation processes associated with zones VI and VII indicate[13] that the critical resolved shear strain for $\{110\}$ twinning (mode $D1_2$) is comparable with that required to cause significant $\{110\}$ slip and is therefore less than that associated with the martensitic transformations. This is confirmed in a later consideration of the deformation behaviour of individual polyethylene spherulites.

The results which have been briefly summarized above show that the following criteria govern in general the operation of deformation modes in chain-folded polyethylene single crystals.

(i) The magnitude of the shear strain associated with a twin or martensitic transformation process must be small.

(ii) The inhomogeneous shuffles of the molecular chains required to produce the final twin or transformation crystal structure and orientation must be simple.

(iii) There must be a simple dislocation mechanism for the deformation process, as indicated by the unique crystallographic features of the operative deformation processes.

(iv) A critical resolved shear strain criterion determines which of all of the recorded deformation modes will be predominant.

(v) Deformation processes which involve a significant extension in fold bend lengths will not operate. Multiple deformation processes will operate if they are consistent with small fold bend extensions and if they individually satisfy the criteria listed above. This implies that extensive deformation by slip will only occur in the plane parallel to the fold plane.

The criteria for the operation of deformation modes which are listed above should apply in general to all synthetic and natural molecular crystals which have a chain-folded configuration and where the deformation is restricted to the transverse displacement of molecular chains. The criteria listed, with the exception of (v), should also apply to molecular crystals which do not possess folded-chain geometry. In all cases, however, the observation of stress-induced martensitic transformation processes will depend critically on criterion (iv), assuming that criteria (i)—(iii) and the tenuously related criterion governing the relative values of the principal strains associated with a transformation can all be satisfied.

Unfortunately, it is not possible to test the validity of the criteria listed above for deformation processes in folded-chain crystals other than polyethylene. This arises because of the lack of detailed experimental evidence about the intra-lamella deformation processes in other polymer crystals owing in part to the difficulty of gaining definite diffraction information from crystalline polymers which can be extremely sensitive to electron beams.[28,29] There is, however, a considerable amount of evidence in the literature which indicates that twinning and martensitic transformations are operative in other polymer crystals. These crystals are worthy of study in order to gain more information about the molecular geometries which are conducive to the operation of twinning and transformations, and to the mechanisms and the effect of fold geometries on the nucleation and propagation of the deformation processes. Polyoxymethylene[9,10] polybutene-1,[6,34] and nylon 66[5] are particularly worthy of study in these respects.

The experimental results on which all of the discussion in this section has been based were obtained from single crystals deformed whilst in contact with a Mylar substrate and therefore following the procedure used by Kiho, Peterlin, and Geil.[3] Recently some experiments[35,36] have been carried out on the deformation of single crystals which were either unsupported during straining or strained whilst in contact with a bulk single crystal substrate to induce straining conditions different from those resulting from Mylar substrate tests. These have shown that chain-axis slip is a very easy mode of deformation and is, in general, preferred to those processes involving only the transverse displacement of molecular chains. These experiments did not, however, result in any further understanding of the mechanisms of twinning and stress-induced martensitic transformations.

In relation to the behaviour of bulk crystalline polymers it is particularly important that the role of martensitic transformations, twinning, intra-lamella slip, chain-axis slip, and inter-lamella deformation processes be identified. It is difficult to identify the roles of these deformation processes using small- and large-angle X-ray diffraction techniques in bulk specimens possessing no preferred orientation. There have, however, been some excellent studies on the deformation of oriented bulk specimens of crystalline polymers[4,37–39] which have provided some definite information about the operation of the deformation processes referred to. Recent studies by Young and Bowden[39] on the deformation of bulk 'chain-extended' 'single' crystals should in particular provide valuable data which will allow a detailed comparison to be made of the deformation behaviour of single crystal and bulk forms of polyethylene.

[34] V. F. Holland and R. L. Miller, *J. Appl. Phys.*, 1964, **32**, 3241.
[35] J. Petermann and H. Gleiter, *J. Polymer Sci., Part A-2, Polymer Phys.*, 1972, **10**, 2333.
[36] W. Wu, A. S. Argon, and A. P. L. Turner, *J. Polymer Sci., Part A-2, Polymer Phys.*, 1972. 10,2397.
[37] F. C. Frank, V. B. Gupta, and I. M. Ward, *Phil. Mag.*, 1970, **21**, 1127.
[38] D. Lewis, E. J. Wheeler, W. F. Maddams, and J. E. Preedy, *J. Appl. Cryst.*, 1971, **4**, 55.
[39] R. J. Young, P. B. Bowden, J. M. Ritchie, and J. G. Rider, *J. Materials Sci.*, 1973, **8**, 23.

In an attempt to identify the roles of the feasible intra- and inter-lamella deformation processes in the deformation of a spherulite, Allan and Bevis[13] have investigated the deformation behaviour of thin, but fully spherulitic films of polyethylene. The films were deformed whilst in contact with a Mylar substrate as described in Section 4. The experimental technique used minimized the effect of reverse stresses due to the existence of the composite amorphous-crystalline structure which can lead to the reversion of the reported deformation processes.[1] Before deformation all regions of the spherulites exhibited a 5μm selected area diffraction pattern of the type shown in Figure 10. All spherulites exhibited the same marked anisotropy of deformation behaviour which was determined by the relative orientations of the spherulite radius of interest and the straining axis. The resultant crystal deformation morphology was most marked at between 30 and 40% deformation as indicated in the electron-micrograph in Figure 21. The banded diffraction contrast shown in Figure 21 occurred when the spherulite radius was approximately normal to the straining axis. The microcracking shown in the same Figure occurred preferentially where the spherulite radius

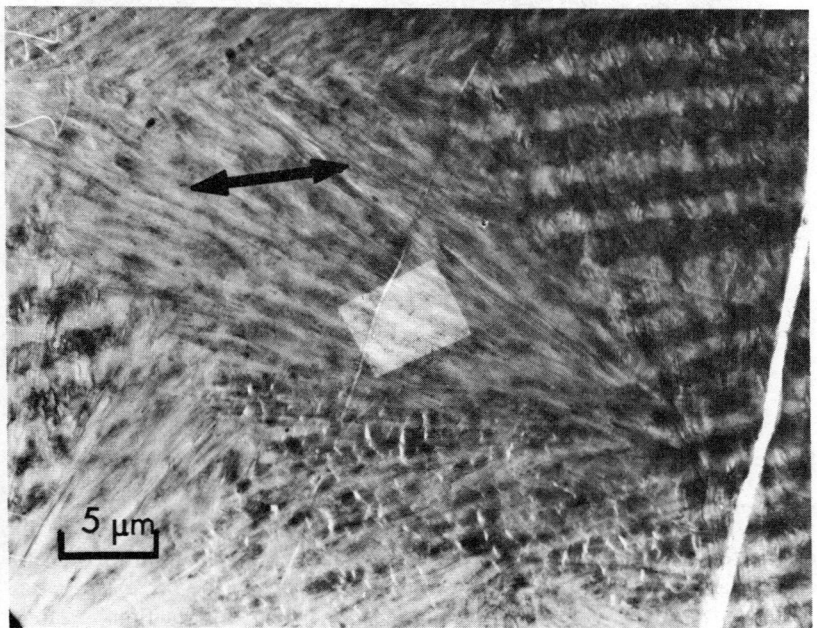

Figure 21 *Transmission electron micrograph showing the marked change in diffraction contrast which occurs within a spherulitic polyethylene film after deformation. The observed contrast effects are determined by the relative orientations of the spherulite radii and the straining axis. The straining axis is indicated by an arrow and is shown to be parallel to the 'banded' diffraction contrast*

was parallel to the straining axis. The intermediate regions exhibited no significant change in morphology with increasing percentage deformation. The electron diffraction patterns obtained from the three regions were different other than at low percentage deformations, and all changed with increasing percentage deformation. Deformation twinning and stress-induced martensitic transformations played a more predominant role in the regions which exhibited the banded diffraction contrast. This region has therefore been selected for discussion here.

The major changes in the diffraction patterns up to 20—30% deformation were due to the formation of twinned diffraction patterns which increase in intensity relative to the intensity of the untwinned (original) patterns with increasing percentage deformation. The patterns obtained[13] represented twinning on both {110} planes of all lamellae orientations about the [010] spherulite radii. This is indicated in the diffraction patterns shown in Figure 22 which was obtained at 40% deformation. In addition to the twin spots,

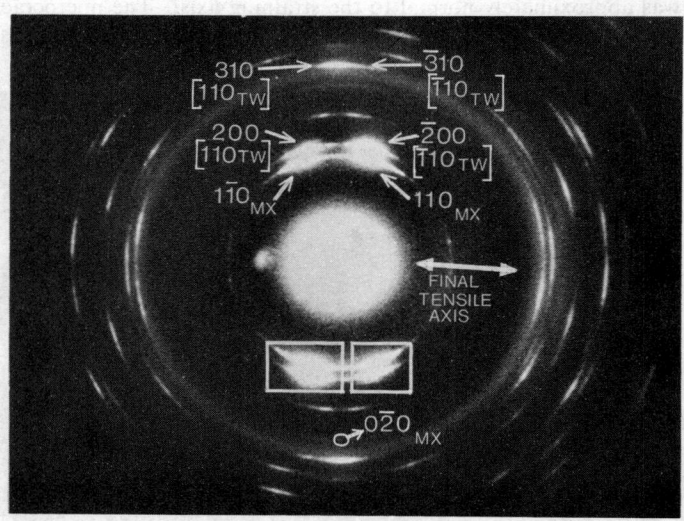

Figure 22 *Selected area electron diffraction pattern of the type obtained from the regions within a spherulite exhibiting a 'banded' diffraction contrast after 40% deformation. The orientation of the final tensile axis is shown*

only a few of which have been indexed, a considerable rotation of the enclosed pairs of diffraction spots has occurred owing to the occurrence of a significant amount of intra- and inter-lamella slip at deformations greater than 30%. The (310) twinned crystal reflections at 40% deformation were very intense. The (310) plane in the twin is approximately parallel to the (010) plane in the matrix (see Figure 9 for the mode $D1_2$) so that the (310) reflection occurs for all twinned volumes. The (310) twin spots increase in intensity at the expense of the (020) matrix spots with increasing deformation.

Scanning-transmission electron microscopy, which allows beam-sensitive materials to be studied at high magnification, has been used to study the relation between the banded diffraction contrast and underlying lamella morphology. Figure 23 is a composite scanning-transmission electron micrograph of a typical set of bands. There were indications that the diffraction contrast was associated with particular underlying lamella crystal orientations, although this has yet to be confirmed.

Reflections from stress-induced martensites were pronounced at deformations greater than 40%.[13] An example of an electron diffraction pattern which exhibited monoclinic reflections and was obtained from spherulite regions exhibiting a banded contrast is shown in Figure 24. The pattern shown was obtained from an area adjacent to the intermediate region, and at smaller percentage deformations exhibited a diffraction pattern which was consistent with twinning on one set of {110} planes. This may be compared with Figure 21 which was consistent with the operation of both sets of {110} twins and was obtained from a region of the spherulite where the spherulite radius was normal to the straining axis. This difference in behaviour may be explained on the basis that, in the latter case, both {110} twin systems were equally stressed. Figure 24 is representative of both matrix and a single set of {110} twins which have undergone a partial $T1_1$ martensitic transformation. The $T1_1$ transformations and the {110} twinning referred to above occurred under straining conditions which were entirely consistent with those observed to govern the operation of the same processes in polyethylene single crystals.

This result is consistent with views expressed by Hay and Keller[5] where twinning in suitably oriented lamellae within a spherulite of nylon 66 was proposed to account for the marked changes in crystalline texture which occurred at relatively low percentage deformations.

The brief summary given above which was based on a selection of results[13] obtained from deformed polyethylene spherulites showed that deformation twinning is an important mode of plastic deformation at small strains. The results indicate that the critical resolved shear strain for twinning is comparable with that of intra-lamella slip and chain-axis slip. It is also clear that the stress-induced martensitic transformations play little part in the early stages of plastic deformation in polyethylene spherulites. The martensitic transformation $T1_1$ was only observed to operate when the spherulites had undergone a substantial elongation and reorientation by other intra- and inter-lamella deformation processes as in the case of the deformation of highly oriented bulk polyethylene.

The electron diffraction and microscopy experiments of the type reviewed above provide a means of refining further the understanding of the role of transformation and twinning processes in the deformation of molecular crystals and of the criteria which govern the operation of these processes. In particular they would be appropriate for the systematic study of deformation processes in molecular crystals other than polyethylene, and also for

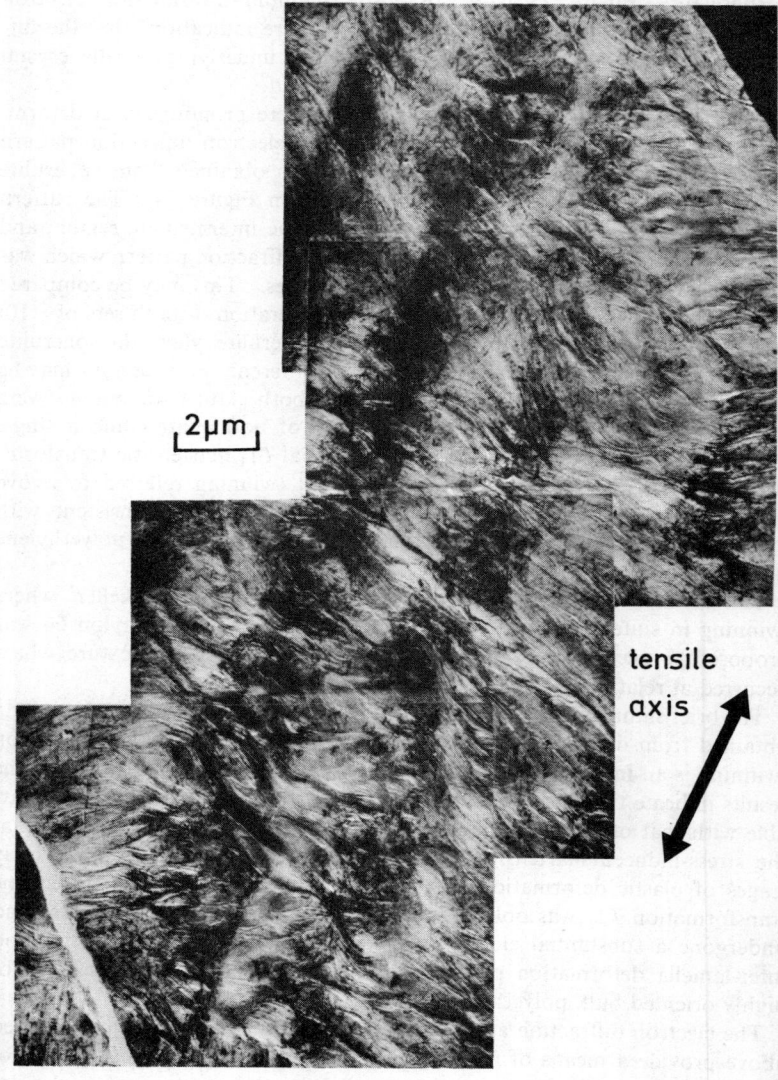

Figure 23 Scanning-transmission electron micrograph of a region of a spherulite exhibiting a 'banded' diffraction contrast.

Martensitic Transformations and Twinning in Organic Crystals

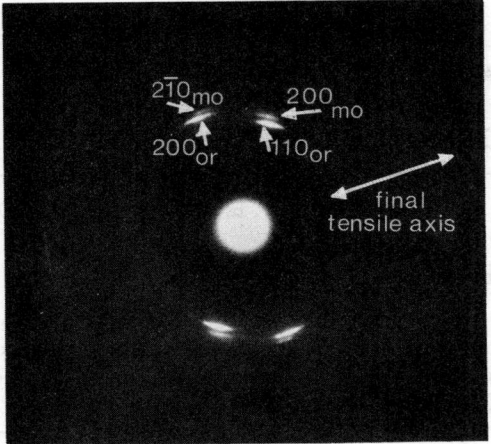

Figure 24 *Selected-area electron diffraction pattern obtained from a region of a spherulite exhibiting a 'banded' diffraction contrast. The pattern represented twinning on just one set of {110} twins due to the significant difference in the relative orientation of the tensile axis and the spherulite radius. The $T1_1$ monoclinic reflections associated with the transformation of both the original and twinned lamellae have been indexed*

the systematic study of the effect of the degree of cross-linking (an extension of recent work by Andrews and Voigt-Martin on the effect of cross-linking on the deformation of polyethylene single crystals[40]) and the density and size of branches on the deformation of spherulites of polyethylene and polyethylene copolymers, respectively.

---

[40] E. H. Andrews and I. G. Voigt-Martin, *Proc. Roy. Soc.*, 1972, **A327**, 251.

4
A Simple Wavefunction for Solid and Surface Calculations

BY R. A. SUTHERS, J. W. LINNETT, AND W. D. ERICKSON

1 Introduction

In this chapter the application of the Floating Spherical Gaussian Orbital (FSGO) method to solids, surfaces, and adsorbed species is described. There are, as yet, few examples of such applications of the method, so that, in some degree, the report provides an indication of how this method might be applied, and value of the results that it would yield. As with all theoretical treatments of an approximate kind, the most important factor is, perhaps, to be able to judge the extent to which the conclusions reached, using the method, will be reliable and useful. An effort will be made to present the authors' views in this respect. However, before the extension of the method to solids and surfaces can be considered, it is necessary to describe the nature of the method and some results for its application to molecules.

The method was introduced by Frost,[1] who was impressed by the level of success achieved by the extremely simple approach of Kimball[2] who assigned electron pairs to spherical, non-overlapping orbitals having a constant magnitude within each sphere and being zero outside. He calculated shapes of molecules with considerable success. His ideas have been extended in a qualitative way by Bent[3] who has, moreover, shown that they can be useful in considering the structure of molecular solids and the way in which the molecules pack together.

The molecular wavefunction proposed by Frost is almost the simplest that can be imagined. In its simple form it is only applicable to molecules or ions for which a consideration of the electrons in pairs (in the sense of G. N. Lewis) provides a good description of the electronic structure. In the form originally used by Frost, each pair is assigned to an orbital consisting of just a single spherical Gaussian function

$$G_i = A_i \cdot \exp[-\alpha_i(\mathbf{r}_i - \mathbf{R}_i)^2]$$

The Gaussian function, G_i, can be described in terms of its 'size' or extension and the position of its centre (Figure 1). The first is dependent on $\alpha_i^{-1/2}$ and

---

[1] A. A. Frost, *J. Chem. Phys.*, 1967, **47**, 3707.
[2] G. E. Kimball (unpublished work). See review by J. R. Platt, in 'Encyclopaedia of Physics', ed. S. Flügge, Springer-Verlag, Berlin, 1961, Part 2, p. 258.
[3] H. A Bent, *J. Chem. Educ.*, 1963, **40**, 446, 523; 1965, **42**, 302, 348.

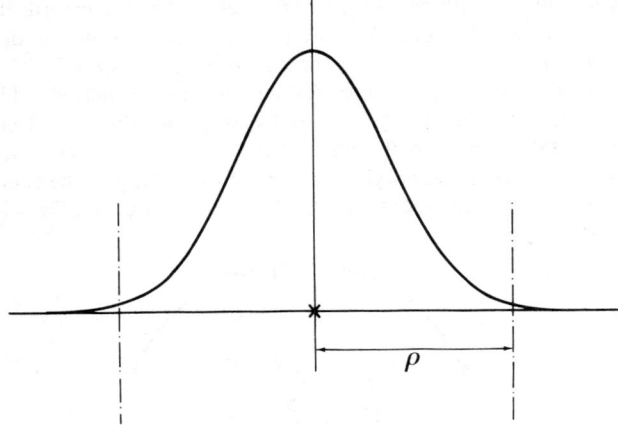

Figure 1 *Cross-sectional plot and 'effective' orbital radius, ρ, of a Gaussian orbital*

the latter is represented by the vector $\mathbf{R}_i$. In the application of the method, these are varied until the 'energy' is a minimum. The 'energy' is computed in the usual way as a weighted average energy by the Variation Method. The Hamiltonian, or quantum mechanical expression of the energy, is exact. That is, it contains all the terms for electron–electron repulsion, electron–nucleus attraction, and nucleus–nucleus repulsion, together with the quantum mechanical operators corresponding to the electronic kinetic energy. So there is no approximation in the Hamiltonian, but only in the extremely simple form used for the wavefunction. Before considering a few simple examples, it must be pointed out here that, when more than one electron pair is present, a single determinantal wavefunction is used. The Pauli Exclusion Principle is therefore satisfied. This is, of course, most important because the shapes and other properties of molecules are determined by the effects of the Principle together with electrostatic effects, the magnitude of which is decided by the magnitude of the charge on an electron, e, and with electronic kinetic energy effects which, in our quantum mechanical world, are determined by an operator, the magnitude being decided by Planck's Constant, h, and the mass of an electron, m. As has been stated above, the form of the wavefunction is extremely simple. Also there is no allowance for the effects of interelectronic repulsion on the electron distribution within each pair. There is some allowance for this effect between pairs in that they can move apart and, in general, they achieve this by 'covering' different parts of space.

The hydrogen molecule will be taken as the first example. It will be described by two protons with two electrons in an orbital described by a single Gaussian function centred (by symmetry) on the mid-point between the two protons. The calculated energy will be dependent on two quantities: the internuclear distance and the value of α for the Gaussian orbital. The energy is minimized with respect to both these. The results are illustrated in Figure 2. It will be seen that the equilibrium internuclear distance is calculated to be 0.78 Å, as against the experimental value of 0.74 Å, which is surely an impressive performance for such a simple function. The energy of a hydrogen atom can be calculated by assigning the one electron to a Gaussian orbital centred on the nucleus (which is, in fact, the location for minimizing the energy) and varying the size of the orbital. The energy of the hydrogen atom is calculated to be 0.4244 a.u. (1 a.u. = 2579 kJ mol^{-1}).

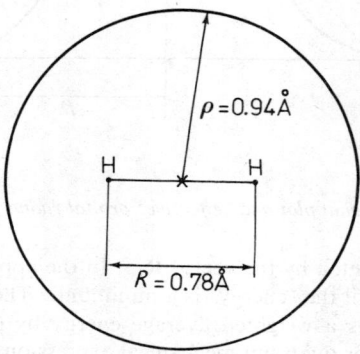

Figure 2 *The FSGO description of the hydrogen molecule.* ($\cdot$ – *nuclear positions;* $\times$ – *orbital centre; orbital radius* $\rho = \alpha^{-1/2}$)

The energy of the hydrogen molecule is calculated to be 0.9559 a.u. so that the dissociation energy for the hydrogen molecule is 0.1071 a.u., which is equal to 276 kJ mol^{-1}. The experimental value is 450 kJ mol^{-1}. On the other hand, it must be remembered that the Hartree–Fock value corresponding to the best MO treatment, in which both the electrons are assigned to the same orbital, is 326 kJ mol^{-1}. The present function could not do better than that and, on this basis, the result must be regarded as satisfactory. In particular, though the absolute errors in the energy for H$_2$ and 2H are considerable, the difference from the Hartree–Fock dissociation energy is much less; and it is, of course, such energy differences which are important to chemists rather than total electronic energies. The error is therefore, in considerable part, the error arising from inadequate allowance for interelectron repulsion. Moreover, this inability to allow adequately for intra-pair electron repulsion is important in this case because the dissociation process separates a pair. Even in the best single orbital treatment (the Hartree–Fock), this separation of the electron pair causes an error of 90 kJ mol^{-1}.

The next example will be the lithium hydride diatomic molecule, LiH. This contains two pairs of electrons and, in the FSGO model, one will be assigned to one Gaussian orbital and the other to a second Gaussian orbital. These two orbitals will be allowed to float in space and size and the separation between the nuclei will be varied. For the lowest energy the separation between the nuclei is found to be 1.71 Å compared with an experimental value of 1.595 Å. It is found that one Gaussian is centred very close to the lithium nucleus and has a small size; this is equivalent to the K-shell orbital. The other is situated on the internuclear line and its centre is fairly close to the proton. The final wavefunction is shown diagrammatically in Figure 3.

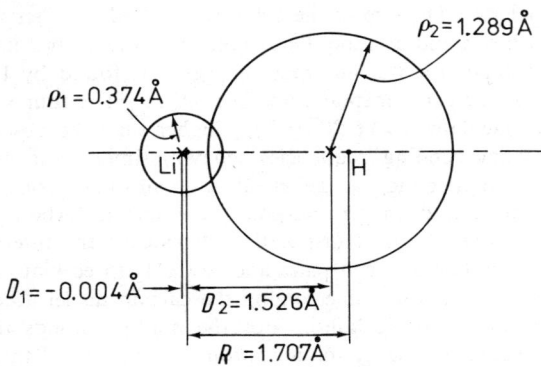

Figure 3 *Gaussian orbitals and nuclear configuration in the FSGO description of the LiH molecule.* ($\cdot$ *– nuclear positions;* × *– orbital centres*)

It will be seen that, in line with the behaviour of LiH, the structure is very ionic, consisting of a lithium positive ion and a polarized negative hydride ion.

The BH molecule contains three pairs and these occupy three Gaussian orbitals; one centred very close to the boron nucleus (the K-shell), another centred between the boron and hydrogen nuclei (the bonding pair), and the third located on the opposite side of the boron nucleus from the proton (the lone pair). The calculated equilibrium bond length is 1.31 against the observed of 1.24 Å. The wavefunction is shown diagrammatically in Figure 4. The vibration frequencies of H_2, LiH, and BH are calculated to be 4743 (4395), 1553 (1406), 2561 (2366), respectively, the experimental values being

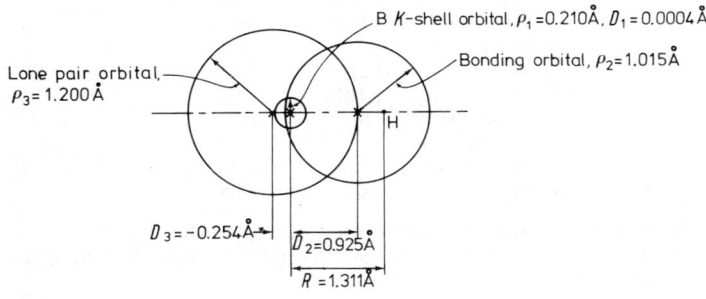

Figure 4 *The FSGO description of the* BH *molecule*

given in brackets (units: cm^{-1}). The computation of energy changes near the minimum of the curve is very satisfactory for these simple examples.

Finally, to exemplify the method, the hydrocarbon molecules CH_4, C_2H_6, C_2H_4, and C_2H_2 will be considered. For methane the model uses a K-shell Gaussian orbital centred on the carbon nucleus (K-shell) and four equivalent Gaussian CH-bonding orbitals tetrahedrally disposed symmetrically round the carbon nucleus. The size of the orbitals is varied together with the distance of the protons and bonding Gaussians from the central nucleus. The internuclear distance for the minimum energy was found by Frost[4] to be 1.115 Å whereas the experimental value is 1.093 Å. The four vibration frequencies were calculated to be 20 to 30% higher than the observed values, both stretching and bending frequencies behaving similarly in this respect.

For C_2H_6 the single bond was described by a pair of electrons occupying a Gaussian orbital centred on the mid-point between the carbon nuclei. For C_2H_4, two Gaussian orbitals disposed each side of the internuclear line accommodated the two electron pairs and in C_2H_2 three Gaussian orbitals disposed trigonally round the line joining the carbon nuclei accommodated the three pairs of the triple bond. For the multiple bonds the bonding Gaussians coalesced on energy minimization, so Frost fixed their centres a small distance from the internuclear line. The results of the calculations are shown in Table 1, which compares the calculated bond lengths and angles with the experimental values for these three molecules. It is seen that the agreement is extraordinarily good. In particular, the FSGO reproduces in form and in magnitude the small decrease in CH bond length that takes place on passing from ethane, through ethylene to acetylene. This is an impressive

Table 1 *Comparison of experimental and FSGO results for the geometrical parameters of ethane, ethylene, and acetylene (Bond lengths quoted in Å)*

	FSGO[a]	Observed[b]
Ethane		
C—C	1.501	1.534
C—H	1.120	1.093
∠H—C—H	108.2°	109.1°
Ethylene		
C=C	1.351	1.337
C—H	1.101	1.086
∠H—C—H	118.7°	117.3°
Acetylene		
C≡C	1.214	1.205
C—H	1.079	1.059

[a] A. A. Frost and R. A. Rouse, *J. Amer. Chem. Soc.*, 1968, **90**, 1965.
[b] 'Interatomic Distances,' ed. L. E. Sutton, (Special Publ. No. 18), The Chemical Society, London, 1965.

[4] A. A. Frost and R. A. Rouse, *J. Amer. Chem Soc.*, 1968, **90**, 1965.

success and shows that the function, presumably because all orbital sizes and positions are allowed to float, is able to reproduce these subtle changes in molecular dimensions.

2 Li_nH_n Cluster Calculation

As an extension of the calculation of Frost for diatomic LiH, some additional calculations have recently been made by Pakiari[5] for Li_2H_2, Li_3H_3, and Li_4H_4. It is hoped that the results will give insight into the condensation of molecules to yield crystals, and that they will provide a valuable link with the results for crystalline LiH.

As previously stated, the internuclear separation for LiH is calculated to be 1.71 Å, compared with the experimental value of 1.60 Å, an error of 7%. For Li_2H_2 the FSGO model predicts an equilibrium conformation of a diamond in which the four nuclei are co-planar. All four LiH internuclear separations have the same value of 1.84 Å, and the LiHLi angle is 74.6°. The total molecular energy of Li_2H_2 is calculated to be -13.2398 a.u. compared with -6.5717 a.u. for LiH. There is, therefore, a lowering of 127 kJ mol^{-1} per LiH unit upon dimerization, which is 58% of the experimental sublimation energy of crystalline LiH. The nearest neighbour internuclear spacing in solid LiH is determined experimentally to be 2.04 Å so the LiH separation in the dimer has also moved some way towards the crystalline value.

A calculation for linear $(LiH)_2$ has also been made, and its total energy was found to be -13.1916 a.u. The energy gained through cyclization is therefore 63 kJ mol^{-1} per LiH unit, which is one half of the heat of dimerization. The equilibrium configurations of the two structures (linear and cyclic) are shown diagrammatically in Figure 5.

A planar hexagonal conformation is predicted for Li_3H_3 in which each of the six LiH sides have the same bond length of 1.79 Å, the LiHLi angles are 101.3°, and the total energy is calculated to be -19.8994 a.u. The energy gain from molecular LiH to the trimer is therefore 161 kJ mol^{-1} per LiH unit, and this is equivalent to 73% of the experimental sublimation energy of crystalline LiH. Linear $(LiH)_3$ has a total energy of -19.8228 a.u., and hence the heat of cyclization is 67 kJ mol^{-1} per LiH unit, which is approximately the same as that for Li_2H_2. The geometrical details of linear $(LiH)_3$ are shown diagrammatically in Figure 6, together with those for cyclic Li_3H_3.

For the Li_4H_4 molecule, a slightly distorted cube is predicted where the Li nuclei are situated at the corners of one regular tetrahedron and the H nuclei at the corners of a second concentric tetrahedron symmetrically disposed relative to the first. The calculated LiH distances are 1.91 Å and the separation of the Li and H nuclei from the common centre of the two tetrahedra are 1.49

---

[5] A. H. Pakiari, private communication.

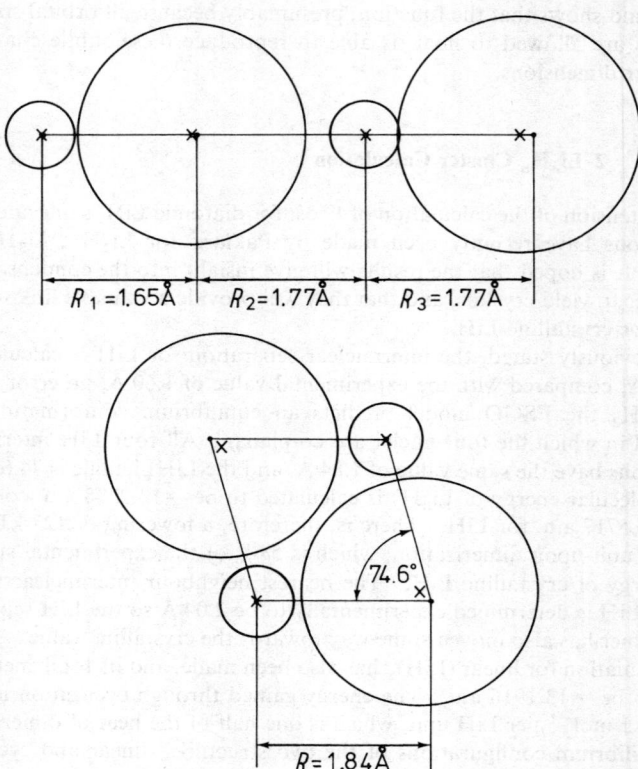

Figure 5 *Linear and cyclic* Li_2H_2. (• – *nuclear positions;* × – *orbital centres*)

and 1.79 Å, respectively. The energy of this tetramer is calculated to be -26.5687 a.u., which is a gain of 185 kJ mol^{-1} per LiH unit relative to diatomic LiH and this is equivalent to 84% of the experimental sublimation energy of crystalline LiH. Putting together the first two LiH molecules therefore yields twice as much energy as putting two dimers together to give the tetramer. Of course, this means that the energy gained in $2\,Li_2H_2 \rightarrow Li_4H_4$ (234 kJ mol^{-1}) or $LiH + Li_3H_3 \rightarrow Li_4H_4$ (256 kJ mol^{-1}) is about the same as that in $2\,LiH \rightarrow Li_2H_2$ (253 kJ mol^{-1}) or $LiH + Li_2H_2 \rightarrow Li_3H_3$ (231 kJ mol^{-1}). However, the interesting result of the present calculations is that a large amount of the crystal binding energy is obtained in the early stages of assembly of these LiH molecules.

Planar Li_4H_4 has a cyclic octahedral conformation in which the LiH sides are 1.77 Å long and the LiHLi angles are 115.8°. Its energy is calculated to be -26.5470 a.u. and hence the energy gain per LiH unit in progressing from

A Simple Wavefunction for Solid and Surface Calculations

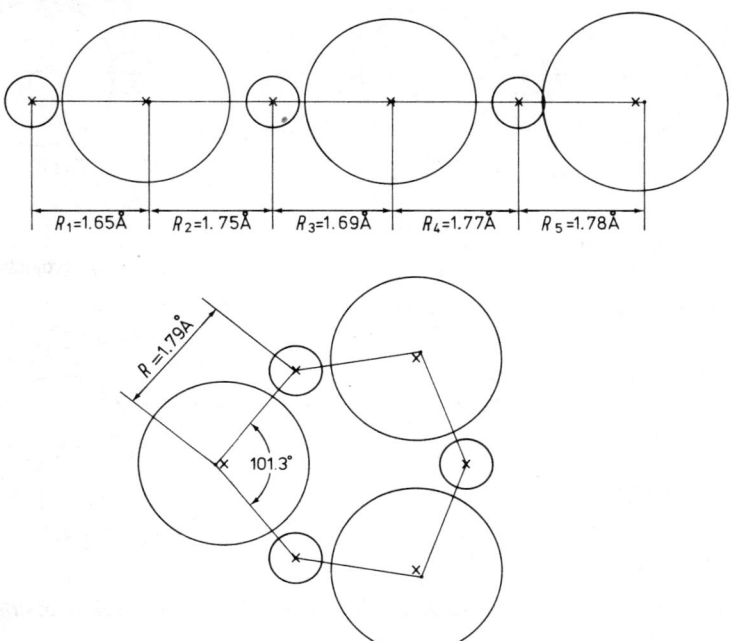

Figure 6 *Linear and cyclic* Li_3H_3. (• – *nuclear positions;* × – *orbital centres*)

a two-dimensional to a three-dimensional conformation is 14 kJ mol^{-1}. It is noted that as a rule the LiH bond lengths shorten with increasing ring size, and this may be explained by decreasing ring strain for cyclic Li_nH_n as n progresses from 2 to 4.

Linear $(LiH)_4$ has a calculated energy of -26.4576 a.u. which gives the heat of cyclization as 59 kJ mol^{-1} per LiH unit. The geometrical structure of $(LiH)_4$ follows a similar trend to those of linear $(LiH)_3$ and $(LiH)_2$ in that the bond lengths alternate short and long with the terminal Li atom giving rise to the shortest internuclear separation. Also, the terminal H$^-$ Gaussian is the only orbital which is polarized to any marked extent within all of the linear $(LiH)_n$ molecules, and hence a similar feature may be expected in the surface region of crystalline LiH. Figure 7 shows diagrammatically the nuclear and Gaussian orbital configurations in the three Li_4H_4 structures discussed above.

Further studies are clearly needed of the energies and shapes of Li_nH_n molecules for $n = 5$—7 *etc.* However, the results so far augur well for the future computations and suggest that they will provide useful indications of the way in which crystallites are formed from the vapour.

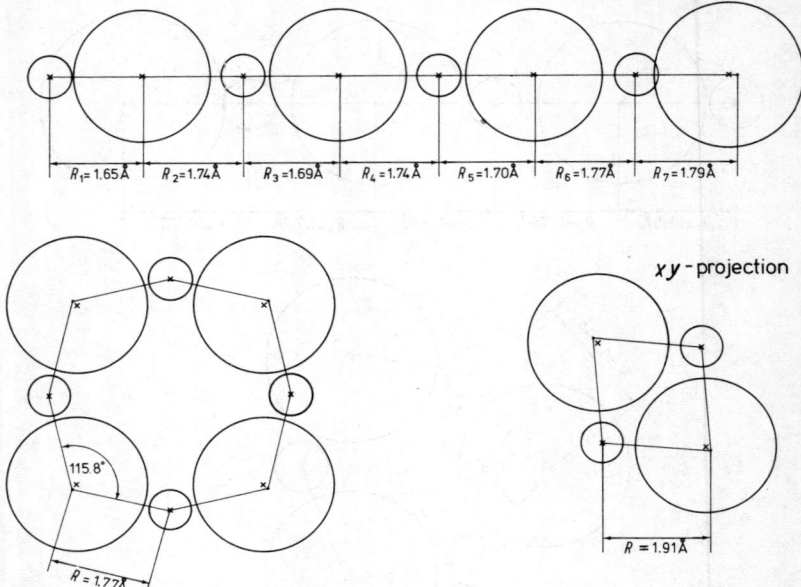

Figure 7 *Linear, planar, and cubic (xy-projection)* Li_4H_4. (• – *nuclear positions;* × – *orbital centres*)

3 The LiH Infinite Crystal

The next logical step after the Li_nH_n cluster calculations is the determination of the electronic and geometrical structure of the LiH infinite crystal. Although the primary aim is an examination of the structural and chemical behaviour of surfaces, it will be appreciated that a definitive calculation on the solid alone is a necessary precursor to any surface calculations.

The evaluation of the crystal energy will clearly involve far more computation than that for the corresponding molecule, simply by the increased number of interactions appearing in the energy expression. For the calculations to remain of a tractable size, a critical study must be made of the crystal energy, and only those terms with a finite contribution will be retained in the energy expression. To this end, the relatively simple LiH crystal is an admirable example upon which to develop a general method which is applicable to any crystalline structure.

Together an Li and H atom contribute four electrons to the total electronic wavefunction, and within the closed-shell approximation these give rise to two doubly occupied, but non-equivalent, basis orbitals. Crystalline LiH has a NaCl-type structure, and hence each Li and H nucleus must be situated in an octahedrally symmetrical electronic environment. This then implies that each of the two types of basis orbital must be centred on an atomic nucleus, either Li or H. One type of Gaussian basis orbital must necessarily be centred on the Li nuclei in order to describe the K-shell electrons of the lithium atoms, but the energetically favoured position of the second type of Gaussian basis orbital is on the H nuclei. The FSGO model therefore imparts an $Li^+ - H^-$ ionic nature to the crystal wavefunction, and this is a logical extension of the LiH molecule results discussed above.

It follows that there are just three parameters in the present method for which the LiH infinite crystal energy will be minimized by the variation principle. These are: (i) α_1, the size of the Gaussian orbitals centred on the Li nuclei; (ii) α_2, the size of the Gaussian orbitals centred on the H nuclei; (iii) R_0, the internuclear spacing between nearest LiH neighbours.

The LiH solid calculation is now achieved by including only those energy terms that are associated with a reference Li site and a reference H site, each of which is symmetrically surrounded by a theoretically infinite number of other lattice sites. A truncation is understandably necessary in the number of interactions over distant lattice sites which are evaluated explicitly, and a number of restrictions are introduced consequent to the following observations.

The values of all the terms appearing in the total energy expression are determined primarily by two factors: (i) the overlap S_{ij} between the Gaussian orbitals G_i and G_j; and (ii) the inverse distance $1/R_{mn}$, where m is one of the reference lattice sites and n is any other lattice site within the crystal. Not all of the terms in the crystal energy expression are dependent on both of these factors. The electronic kinetic energy is essentially a function of (i) only, and the nucleus–nucleus repulsion is a function of (ii) only. However, the electron–nucleus attraction and the electron–electron repulsion are dependent on both factors.

Simplifications in the crystal energy can therefore be made by determining: (i) those basis orbitals which have a finite overlap with the reference Li and H Gaussian orbitals; and (ii) the maximum distance R_{mn} (and hence the maximum number of lattice sites) beyond which there is a cancellation of the effects of the electron–nucleus attraction with the electron–electron and nucleus–nucleus repulsions.

Consider first the question of finite overlap. For convenience the lattice distances, r, will be measured in terms of normalized internuclear distances $\bar{r} = r/R_0$; Table 2 lists the number of sites in successive sub-lattices which result from symmetrically adding next nearest neighbours at distances $\bar{r} = 1, \sqrt{2}, \sqrt{3}, \sqrt{4}, \ldots$, to the reference LiH unit.

Numerical values for S_{ij} over a wide range of $\bar{r}$ have been reported by Erickson and Linnett,[6] and a typical set of results is reproduced in Table 3. The overlap decreases exponentially as the square of the interorbital distance, and an overlap larger than 0.001 generally only occurs for neighbours which are $\bar{r} = 1$ internuclear units apart for LiH pairs, and $\bar{r} = \sqrt{2}$ and 2 internuclear units apart for HH pairs. To a fair approximation, overlaps of less than

---

[6] W. D. Erickson and J. W. Linnett, *J.C.S. Faraday II*, 1972, **68**, 693.

Table 2 *Total number of sites in sub-lattices arranged symmetrically about the reference LiH unit*

Minimum distance $\bar{r}$, in units of R_0, to which sites about reference LiH unit are included	Total number of sites, N, in sub-lattice
1	12
$\sqrt{2}$	28
$\sqrt{3}$	36
2	46
$\sqrt{5}$	78
$\sqrt{6}$	102
$\sqrt{8}$	118
3	152
$\sqrt{10}$	184

Table 3 *Typical numerical values of S_{ij} and T_{ij} for a $N = 46$ site sub-lattice of Li^+—H^- ions ($\alpha_1 = 2.0$, $\alpha_2 = 0.2$, $R_0 = 3.7$ a.u.)*

Overlap	$\bar{r}$	S_{ij}	T_{ij}
H^-—H^-	0	1.000	1.048
	$\sqrt{2}$	6.470 (−2)	−5.378 (−2)
	2	4.186 (−3)	8.962 (−3)
	$\sqrt{6}$	2.708 (−4)	5.197 (−3)
Li^+—Li^+	0	1.000	1.007
	$\sqrt{2}$	1.285 (−12)	1.752 (−3)
Li^+—H^-	1	3.618 (−2)	−3.040 (−2)
	$\sqrt{3}$	2.492 (−4)	5.088 (−3)

10^{-3} in magnitude may be assumed to be negligible, and therefore S_{ij} is set to zero for all the remaining internuclear separations.

As an aside, it may be noted that the product $S_{ij}T_{ij}$ provides a better criterion by which to judge the value of any omitted terms from the total energy expression. Here the quantities T_{ij} are the inverse overlap matrix elements (*i.e.* $T = S^{-1}$) and arise from the orthogonalization of the Gaussian orbitals, a necessary step in the evaluation of the electronic wavefunction. Appropriate T_{ij} values are listed in Table 3, and it may be seen that all interactions with an $S_{ij}T_{ij}$ value greater than 10^{-5} are included in the energy calculation.

It now remains to determine the effective size of an infinite crystal. This is achieved by using a finite lattice arrangement but in conjunction with the Evjen[7] method of summation. The reference LiH unit is symmetrically surrounded by additional LiH pairs so as to form a rectangular block of dimension $[(2n-1) \times (2n-1) \times 2n]$ where $n = 2,3,4,5$, *etc.* According to the Evjen scheme, all centres of charge in the interior of the block are assigned

[7] H. M. Evjen, *Phys. Rev.*, 1932, **39**, 675.

a unit weight, while those on a face, an edge, and a corner are assigned weights of $\frac{1}{2}$, $\frac{1}{4}$, and $\frac{1}{8}$, respectively.

Calculations for both a $5 \times 5 \times 6$ and a $7 \times 7 \times 8$ block of atoms have been made by Erickson and Linnett,[8] and their results are reproduced in Table 4.

Table 4 *Numerical results for molecular and bulk crystalline* LiH

	Molecule	Bulk crystal	
Block size	$1 \times 1 \times 2$	$5 \times 5 \times 6$	$7 \times 7 \times 8$
Optimized parameters			
α_1	1.998	1.996	1.996
α_2	0.169	0.204	0.204
R_0/a.u.	3.226	3.803	3.805
Total energy for reference LiH unit			
E/a.u.	−6.5727	−6.6731	−6.6725
Calculated heat of sublimation			
ΔH_{subl}/kJ mol^{-1}		263.7	261.9

The two computed LiH internuclear distances for the crystal compare well with the experimental value[9] of 3.861 a.u. (1 a.u. = 0.5292 Å) and both are only some 1.5% too short. In addition, an estimate for the heat of sublimation may be obtained from the equation

$$\Delta H_{\text{subl}} = E(\text{g}) - E(\text{s})$$

Values just larger than 260 kJ mol^{-1} are obtained, and these again compare well with the experimental value[10] of 220 kJ mol^{-1}.

It is noted that little variation occurs in either the optimized parameter values or the calculated total crystal energy in progressing from a rectangular block of $5 \times 5 \times 6$ atoms to one of $7 \times 7 \times 8$ atoms. However, the variations that do occur are in the correct direction; *viz.*, both the internuclear spacing and the heat of sublimation tend toward their experimental values. Therefore it is concluded that a $5 \times 5 \times 6$ block of atoms is a sufficient lattice size to simulate an infinite crystal, and that an extension of the above scheme to surface calculations is expected to produce reasonable results for the lattice dimensions in the region of the surface.

4 The [100] Surface of Crystalline LiH

Continuing from these infinite crystal calculations, a study is now made of the [100] surface of crystalline LiH. Fortunately, many of the modifications introduced to the FSGO method for the infinite crystal calculations are equally

[8] W. D. Erickson and J. W. Linnett, *Proc. Roy. Soc.*, 1972, **A331**, 347.
[9] R. Starizky and D. Walker, *Analyt. Chem.*, 1956, **28**, 1055.
[10] S. R. Gunn and L. G. Green, *J. Amer. Chem. Soc.*, 1958, **80**, 4782.

applicable to the calculation of surfaces. Thus the same limitations of finite overlap, and the same 5 × 5 × 6 block of atoms will be adopted without discussion. There are, however, a number of additional modifications which require to be made because of the special effects associated with a surface.

One of the 5 × 6 sides of the LiH block will be taken to represent the [100] surface, and the remaining five sides will be assumed to border on to an infinite crystal. Consequently, the Evjen method of summation requires a somewhat different mode of application in the surface calculations than was used for the infinite crystal. For the centres of charge not situated on the surface layer, the usual weights of $\frac{1}{2}$, $\frac{1}{4}$, and $\frac{1}{8}$ are assigned to those centres on a face, an edge and a corner. However, for the sites on the surface layer, weights of unit, $\frac{1}{2}$, and $\frac{1}{4}$ are assigned to the centres of charge situated on the face, an edge, and a corner, respectively.

The perturbing influence of the surface on the total lattice energy is clearly not confined to the final [100] layer of atoms, and the energy minimization will require the evaluation of the local energy associated with successive layers of the lattice. Reference LiH units are therefore defined for every layer of the lattice, although only those LiH pairs with an energy different from the bulk crystal value need to be considered explicitly. The reference units are taken to be the central Li and H atoms in each layer, and thus the LiH pairs are aligned one under the other throughout the lattice. Erickson and Linnett[8] have concluded that only those energies associated with the first three surface layers vary appreciably from the infinite crystal value, and hence the sum of these three terms is minimized by the variation principle.

As in the bulk crystal calculations, the energy minimization is made with respect to the orbital and nuclear positions and the orbital sizes. Within the [100] surface calculations the lattice spacings parallel to the surface are fixed at the value obtained for the infinite crystal. However, in the direction perpendicular to the surface, the nuclei and orbitals have a degree of freedom that was not permitted in the bulk crystal, and consequently the orbital and nuclear centres are not necessarily coincident. Associated with each lattice layer there are now six parameters which require to be optimized, $viz.$: α_1 and α_2, the sizes of the two types of Gaussian orbitals; D_1 and D_2, the vertical positions of the Gaussian orbitals; R_1 and R_2, the vertical positions of the Li and H nuclei.

Calculations have been made by Erickson and Linnett[8] in which optimized parameters for the first two surface layers have been obtained. Results for the lattice configuration in the region of the surface are displayed diagrammatically in Figure 8. The distance from the Li nuclei on the surface layer to the nearest neighbour H nuclei in the penultimate layer was calculated to be 0.032 Å shorter than the corresponding crystal value of $R_0 = 2.016$ Å, whereas the distance from the H nuclei on the surface to the nearest Li nuclei in the penultimate layer is 0.158 Å longer than R_0. This outward displacement of the negative ion and inward displacement of the positive ion (*i.e.*, in this case

A Simple Wavefunction for Solid and Surface Calculations

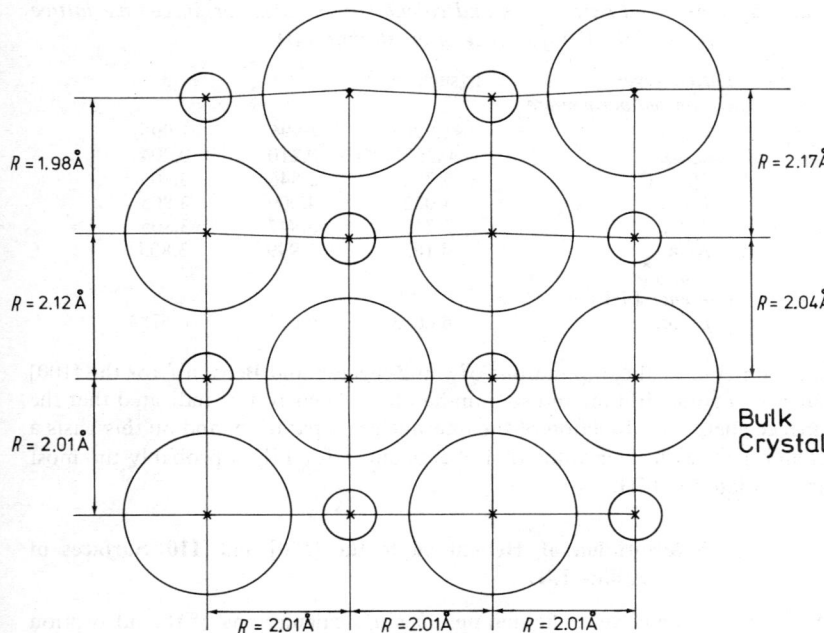

Figure 8 *Lattice arrangement for the* [100] *surface of crystalline* LiH. (• – *nuclear positions*; × – *orbital centres*)

the H and Li atoms, respectively) has also been noted by Verwey[11] in some semi-empirical calculations for a number of alkali halide crystals.

The orbital centre of the surface lithium atom K-shell electrons is understandably only slightly displaced (0.001 Å) from the nuclear position, but an appreciable displacement of 0.041 Å is calculated for the H$^-$ ion pair of electrons and this indicates a move towards the diatomic molecule structure. The polarization of the Li and H ions in the penultimate layer is considerably less than the corresponding effects described above, and hence the assumption that the nuclear and orbital centres are coincident for the third and subsequent layers appears to be a reasonable approximation.

Table 5 lists the energies calculated by Erickson and Linnett[8] for successive layers of the [100] surface of crystalline LiH. An estimate of the surface energy may be obtained from these results by adding the differences between the successive layer energies and the infinite crystal energy, *i.e.*, (0.0068 + 0.0020 − 0.0052) = 0.0036 a.u. The surface area associated with each LiH unit is $2R_0^2$ or 8.13 Å^2, and hence the energy of the [100] surface is calculated to be 193 ergs cm^{-2}. This result appears to be in agreement with

[11] E. J. W. Verwey, *Rec. Trav. chim.*, 1946, **65**, 521.

Table 5 *Optimized parameters and calculated energies for successive lattice layers of the* [100] *surface of crystalline* LiH

Lattice layer	1 (surface)	2	3
Optimized parameters			
α_1	1.996	1.996	1.996
α_2	0.192	0.210	0.204
D_1/a.u.	3.742	3.846	3.805
D_2/a.u.	4.022	4.009	3.805
R_1/a.u.	3.742	3.847	3.805
R_2/a.u.	4.102	3.999	3.805
Energy for reference LiH unit			
E/a.u.	-6.6663	-6.6711	-6.6783

the semi-empirical computations of van Zeggeren and Benson[12] for the [100] surface of some lithium and sodium halides. Their results indicated that the surface energy is a function of the internuclear separation, and on this basis a comparison with their value of 169 ergs cm^{-2} for LiF is probably the most appropriate for LiH.

5 Adsorption of Helium on to the [100] and [110] Surfaces of crystalline LiH.

A series of simple, yet enlightening, *ab initio* calculations of the adsorption of a helium atom on the [100] and [110] surfaces of crystalline LiH have recently been made by Wood and Linnett.[13] Although the ultimate aim is a theoretical study of chemisorption, a great deal of confidence may be placed in the FSGO model if it describes satisfactorily the quantitatively smaller physical adsorption normally associated with the inert gases.

Taking as the starting point the previously described LiH surface calculations, a number of additional simplifications are introduced in the light of the results obtained by Erickson and Linnett.[8] Hence, just the first two surface layers are deemed active in the adsorption process, with the third and successive lattice layers having little or no effect upon the results.

A similar size surface area of LiH atoms is used for the adsorption studies as was used in the [100] surface calculations, but with the added proviso that the He atom is situated in a symmetrical environment. Thus a 5 × 5 surface area of atoms is taken to be sufficient for adsorption on to a central lattice site or vacancy, but a 5 × 6 and a 6 × 6 surface area of atoms is necessary for adsorption on to an atom–atom mid-point or the centre of a square of atoms, respectively.

A helium atom is described quite simply in the FSGO method by a nucleus of charge $+2$ and a single, doubly occupied, Gaussian orbital. For the adsorbate there are, therefore, three parameters which are to be optimized

---
[12] F. van Zeggeren and G. C. Benson, *J. Chem. Phys.*, 1957, **26**, 1077.
[13] J. C. Wood and J. W. Linnett, to be published.

within the energy minimization procedure, and these are: α_{He}, the size of the helium Gaussian orbital; D_{He}, the vertical position above the LiH surface of the Gaussian orbital; R_{He}, the vertical position above the LiH surface of the nucleus.

Although the electronic and geometrical configurations of the lattice ions have been calculated for the [100] surface of LiH, to date no similar calculation has been made for the [110] surface. Consequently, a standardization in the lattice parameters is necessary if the two sets of adsorption results are to be comparable. The lattice ions in each surface layer are therefore taken to be co-planar, and the nearest neighbour LiH internuclear separation is set to the experimental value of 2.04 Å. In addition, the Gaussian orbital sizes are taken to be those calculated by Erickson and Linnett[8] for the infinite LiH crystal, and the orbital and nuclear centres are constrained to be coincident. Adsorption calculations have been made in which the [100] surface parameters of Erickson and Linnett were used, but these vary little from the bulk crystal parameters and have a negligible effect upon the results. Optimized surface parameters within the presence of a helium atom have also been obtained in a number of test calculations, and again little change was observed in the results.

A number of different adsorption sites have been investigated for both the [100] and the [110] surfaces of crystalline LiH, and the results of these calculations are shown graphically in Figures 9 and 10. For the [100] surface, the Li^+ and H^- lattice ions, together with the centre of the square of ions, are continually repulsive towards a helium atom, and only the Li^+ and H^- vacancies provide suitable sites for adsorption to occur. Figure 11 shows diagrammatically the most favoured site of adsorption for He, namely above a H^- vacancy, for which the heat of adsorption is 7.1 kJ mol^{-1}. It is noted that the He atom does not lie in the plane of the surface atoms but is adsorbed at 1.2 Å above the H^- vacancy site. In the case of the Li^+ vacancy, the equilibrium adsorbate position is calculated to be still further away from the surface at 2.2 Å, and consequently the heat of adsorption is reduced to 3.4 kJ mol^{-1}.

In order to determine the physical reasons for the helium atom adsorption, additional calculations have been made in which the various contributions to the heat of adsorption have been evaluated. For the [100] H^- vacancy calculations, the interaction of the He atom with the Li^+ lattice ions produces an attractive energy of 15.4 kJ mol^{-1}, whereas the interaction of the He atom with the H^- lattice ions produces a repulsive energy of -16.2 kJ mol^{-1}. These two contributions have a near cancelling effect, and the heat of adsorption (7.1 kJ mol^{-1}) is therefore determined primarily in this instance by the attractive interaction (7.9 kJ mol^{-1}) of the He atom with the Li^+-H^- overlap charge densities.

A similar effect is observed in the [100] Li^+ vacancy calculations, but with the interactions between the He atom and the lattice ions acting in a contrawise manner. Thus the interactions between the He atom and the H^- lattice

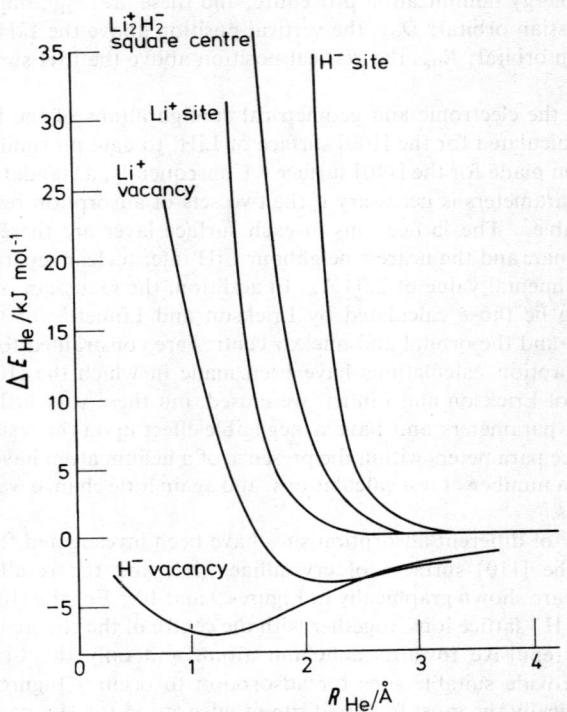

Figure 9 Change in helium atom energy, ΔE_{He}, as a function of height, R_{He}, above the [100] surface of crystalline LiH.

ions now provide a binding energy of 12.2 kJ mol^{-1}, while the interactions between the He atom and the Li$^+$ lattice ions provide a repulsive energy of -13.4 kJ mol^{-1}. Again these two contributions have a near cancelling effect, and the binding energy of 4.6 kJ mol^{-1} obtained for the interaction of the He atom with the Li$^+$—H$^-$ overlap charge densities is an important contributor to the overall heat of adsorption of 3.4 kJ mol^{-1}.

An explanation of the above results is to be found in the ion-induced-dipole electrostatic forces which are automatically included in the FSGO method by the 'floating' of the Gaussian orbitals. Hence, when the nearest lattice neighbour to the adsorbate helium atom is an Li$^+$ ion, the He orbital centre is shifted slightly towards the surface and away from the helium nucleus (thus, for the [100] H$^-$ vacancy calculations; $D_{He} = 1.175$ Å, $R_{He} = 1.191$ Å). Conversely, when an H$^-$ ion is closest to the helium atom, the He orbital centre is situated on the far side of the helium nucleus, away from the surface ions (e.g., for the [100] Li$^+$ vacancy calculations; $D_{He} = 2.180$ Å, $R_{He} = 2.170$ Å). This dependence of the polarization of the He atom upon the nature of

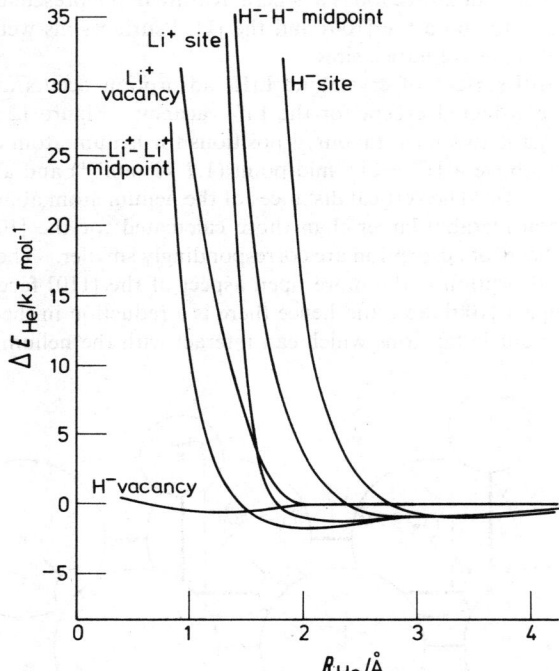

Figure 10 *Change in helium atom energy, ΔE_{He}, as a function of height, R_{He}, above the [110] surface of crystalline LiH*

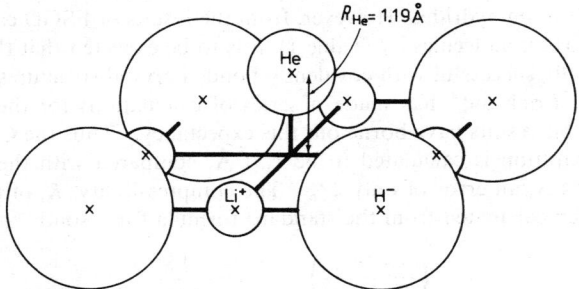

Figure 11 *Adsorption of a He atom on to a H^- vacancy of the [100] surface of crystalline LiH.*

the nearest neighbour lattice ion is a general feature in the present calculations and is observed for both the [100] and the [110] surfaces, as well as for the attractive and repulsive lattice sites.

For the [110] surface of crystalline LiH, adsorption occurs at all of the surface sites considered except for the Li^+ vacancy. Figure 12 shows diagrammatically the two most favoured positions for helium atom adsorption, and these are above a Li^+-Li^+ mid-point (1.7 kJ mol^{-1}) and above a Li^+ ion (1.3 kJ mol^{-1}). The vertical distances of the helium atom above the [110] surface are considerably larger than those calculated for the [100] surface, and thus the heats of adsorption are correspondingly smaller. The reason for this weaker adsorption is the more open aspect of the [110] face compared with the compact [100] face, and hence there is a reduction in the number of nearest neighbour lattice ions which can interact with the helium atom.

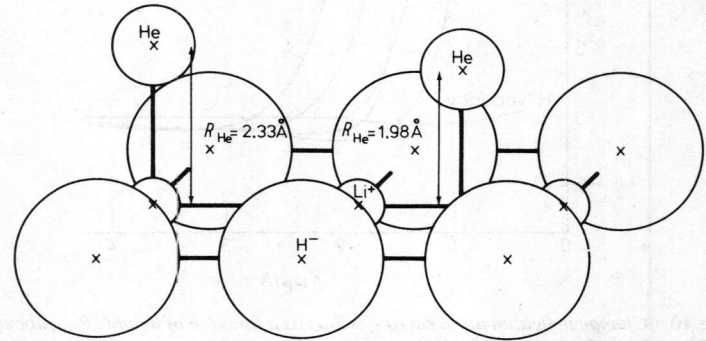

Figure 12 *Adsorption of a* He *atom on to a* Li^+ *site and above the* Li^+-Li^+ *midpoint of the* [110] *surface of crystalline* LiH

6 Conclusion

So far the FSGO method has been applied to the purely ionic structure of crystalline lithium hydride. However, from the results of FSGO calculations for hydrocarbon molecules (*cf.* Table 1), it is to be expected that the method will be equally successful with covalently bonded crystal structures.

Recently, Erickson[14] has made a series of calculations for the diamond crystal and his results have borne out this expectancy. Thus the C—C internuclear separation is calculated to be 1.51 Å, compared with the observed value of 1.54 Å, an error of only 2%. The compressibility, K, of the crystal has also been calculated from the standard formula for a solid

$$\left(\frac{\partial^2 E}{\partial (\log V)^2}\right)_{V=V_0} = -\frac{V_0}{K}$$

---

[14] W. D. Erickson, to be published.

which in terms of the C—C internuclear spacing R_0 becomes

$$\frac{1}{8\sqrt{3}\,R_0}\left(\frac{\partial^2 E}{\partial R^2}\right)_{R=R_0} = -\frac{1}{K}$$

With the value of $R_0 = 1.51$ Å, $(\partial^2 E/\partial R^2)$ is calculated to be 16.8 mdyn Å^{-1}, and the compressibility K is then calculated to be 0.125×10^{-12} cm^2 dyn^{-1}. The experimental value[15] of K is 0.18×10^{-12} cm^2 dyn^{-1}, and hence there is an error of 30% in the calculated value. Since there is little error in the predicted value of R_0, the majority of the error in K is due to the term $(\partial^2 E/\partial R^2)$. It is noted, however, that $(\partial^2 E/\partial R^2)$ is also the definition of a bond-stretching force constant for a diatomic molecule, and these in their turn are generally predicted to be around 30% too large when calculated by the FSGO method.

The energy of a single carbon atom within the diamond crystal is calculated to be -32.0451 a.u., and hence in conjunction with the molecule results of Frost,[4] it is now possible to make estimates for the heats of formation of saturated hydrocarbons such as methane and ethane. Table 6 lists the cal-

Table 6 *Calculated and experimental heats of formation for methane and ethane*

FSGO calculated atoms and molecules	Calculated energy kJ mol^{-1}	Experimental value kJ mol^{-1}
C (diamond)	$-84\,136$	
H$_2$ (g)	-2510	
CH$_4$ (g)	$-89\,248$	
C$_2$H$_6$ (g)	$-175\,925$	
ΔH_f (methane)	-92.3	-76.4
ΔH_f (ethane)	-122.8	-88.5

culated values of ΔH_f for these two molecules, together with the experimental values[16] which have been corrected by C(graphite) → C(diamond) = 1.9 kJ mol^{-1}. The results are in error by 21 and 39% for methane and ethane, respectively, which again demonstrates that energy differences are well reproduced by the FSGO method, even though the absolute energies are far from being exact.

To conclude, calculations are in progress for lithium fluoride, and the possibility of making calculations on the zincite arrangement of BeO is also being investigated. It is felt, however, that both of these crystal studies will lead to meaningful results within the FSGO method, and that this also applies to any chemisorption calculations which may be made.

[15] L. H. Adams and E. D. Williamson, *J. Frank. Inst.*, 1923, **195**, 493.
[16] G. W. C. Kaye and T. H. Laby, 'Tables of Physical and Chemical Constants', Longmans, London, 1959.

7 Bibliography of FSGO Molecule Calculations

A. A. Frost, 'Floating Spherical Gaussian Orbital Model of Molecular Structure. I. Computational Procedure. LiH as an Example', *J. Chem. Phys.*, 1967, **47**, 3707.

A. A. Frost, 'II. One- and Two-Electron-Pair Systems', *J. Chem. Phys.*, 1967, **47**, 3714.

A. A. Frost, 'III. First-Row Atom Hydrides', *J. Phys. Chem.*, 1968, **72**, 1289.

A. A. Frost and R. A. Rouse, 'IV. Hydrocarbons', *J. Amer. Chem. Soc.*, 1968, **90**, 1965.

A. A. Frost, R. A. Rouse, and L. Vescelius: 'V. Computer Programs', *Internat. J. Quantum Chem.*, 1968, **25**, 43.

R. A. Rouse and A. A. Frost, 'VI. Double-Gaussian Modification', *J. Chem. Phys.*, 1969, **50**, 1705.

A. A. Frost, 'VII. Borazine and Diborane', *Theor. chim. Acta*, 1970, **18**, 156.

S. Y. Chu and A. A. Frost, 'VIII. Second-Row Atom Hydrides', *J. Chem. Phys.*, 1971, **54**, 760.

S. Y. Chu and A. A. Frost, 'IX. Diatomic Molecules of First-Row and Second-Row Atoms', *J. Chem. Phys.*, 1971, **54**, 764.

J. L. Nelson and A. A. Frost, 'X. C_3 and C_4 Saturated Hydrocarbons and Cyclobutane', *J. Amer. Chem. Soc.*, 1972, **94**, 3727.

J. L. Nelson and A. A. Frost, 'ESCA Chemical Shifts for Inner Shell Electrons for Small Hydrocarbons', *Chem. Phys. Letters*, 1972, **13**, 610.

J. L. Nelson and A. A. Frost, 'Local Orbitals for Bonding in Ethane', *Theor. chim. Acta*, 1973, **29**, 75.

B. Ford, G. G. Hall, and J. C. Packer, 'Molecular Modelling with Spherical Gaussians', *Internat. J. Quantum Chem.*, 1970, **4**, 533.

P. Th. van Duijnen and D. B. Cook, '*Ab initio* calculations with small ellipsoidal gaussian basis sets', *Mol. Phys.*, 1971, **21**, 475.

R. E. Christoffersen and G. M. Maggiora, '*Ab initio* Calculations on Large Molecules using Molecular Fragments. Preliminary Investigations', *Chem. Phys. Letters*, 1969, **3**, 419.

R. E. Christoffersen, '*Ab initio* Calculations on Large Molecules', *Adv. Quant. Chem.*, 1972, **6**, 333.

L. P. Tan and J. W. Linnett, 'The Lone-pair Orbital in NH_3 and the Calculation of the HNH Angle', *J.C.S Chem. Comm.*, 1973, 737.

5
Appearance Potential Spectroscopy and Related Techniques

BY A. M. BRADSHAW

1 Introduction

During the past four years, the recently revived technique of critical potential determination—retitled soft X-ray appearance potential spectroscopy (SXAPS, or more simply APS)—has aroused great interest within the field of surface studies.[1,2] It is without any doubt the simplest technique available for measuring a set of core electron binding energies and gives in addition valuable density of states information. The application of the method permits a determination of the threshold potential for the appearance of characteristic soft X-rays. Briefly, the method is used as follows: monochromatic electrons are accelerated on to a solid sample giving rise to Bremsstrahlung and characteristic X-ray lines. If the energy of the incident electrons is scanned upward, it is found that small postive steps in the total X-ray yield from the sample occur at energies just large enough to excite core electrons to the Fermi level. Such a step thus corresponds to the onset of a series of characteristic lines. The photon yield is converted into an electron current by means of a photocathode. Electronic differentiation techniques are generally used to enhance the intensity of the step-like features relative to the background, producing a spectrum consisting essentially of peaks (see Figure 1). The intensity of an APS feature at the threshold energy and above will depend on the core hole excitation rate at that energy and its shape is independent of the de-excitation process. The final state in the excitation will consist of a core hole, an excited core electron *and* the scattered incident electron. As will be shown later, the distribution of energy between the two slow electrons must be considered, which leads to the self-convolution of the density of empty states appearing in the expression for the excitation rate. That the shape of an APS feature is independent of the de-excitation mechanism is due to the non-dispersive nature of the technique. The core level binding energies are determined solely by the exciting agent—the incident electron. For each core hole created there is a fixed probability of decay by photon emission (termed the fluorescence yield, ω) which is independent of the energy of the incident electron. Since neither an electron energy analyser as used in X-ray photoelectron spectroscopy (XPS), nor a

---
[1] R. L. Park, J. E. Houston, and D. G. Schriener, *Rev. Sci. Instr.*, 1970, **41**, 1810.
[2] R. L. Park and J. E. Houston, *J. Vac. Sci. Technol.*, 1971, **8**, 91.

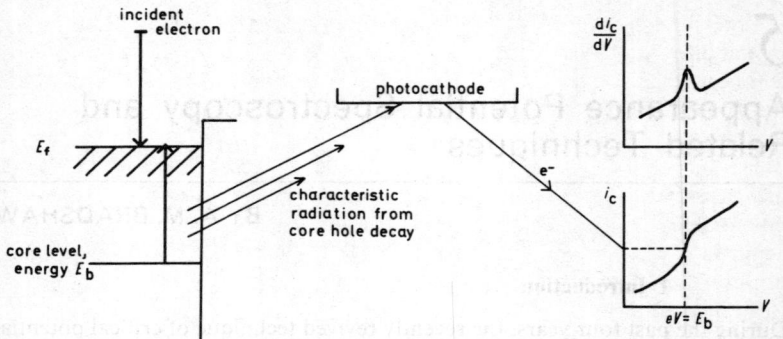

Figure 1 *Schematic representation of the soft X-ray appearance potential experiment*

monochromator as used in X-ray absorption spectroscopy, is required, APS is undoubtedly the simplest method of determining binding energies.

Incident electrons involved in additional inelastic scattering processes will in general not contribute to the peak intensity at or near the threshold energy. The mean penetration depth for elastically scattered electrons in the solid, which may be defined as the mean penetration depth without suffering an energy loss, is given by the elastic mean free path, $\lambda(E)$. As the mean free path of the emitted photons can be taken as being an order of magnitude greater, the average excitation depth in APS is also given by $\lambda(E)$. Most recent determinations of $\lambda(E)$ put its value at 5—20 Å for E within the range 5—1500 eV. Thus in surface chemical analysis SXAPS would appear to offer the same high surface sensitivity as Auger and XPE spectroscopy.

The measurement of 'critical potentials' between 1920 and 1935 was interestingly enough carried out using essentially the same apparatus as the modern APS measurements. A bibliography of the literature during this period is contained in a book by Bruining.[3] The large number of features in the spectra published in the interval to 1930 suggested at first a contradiction of Moseley's results and of the Bohr model of the atom. Later authors tended to ascribe the multitudinous steps and discontinuities in the simple yield curve as being due to diffraction effects in the sample.[4,5] It did not appear at the time that much of the structure could have been due to contaminated surfaces, resulting from inadequate preparation techniques and poor vacuum conditions. An exception is provided by the work of Skinner on lithium[4] and beryllium.[6] He took special precautions to generate clean

---

[3] H. Bruining, 'Physics and Applications of Secondary Electron Emission', Pergamon Press, London, 1954.
[4] H. W. B. Skinner, *Proc. Roy. Soc.*, 1932, **A135**, 84.
[5] M. L. Williams, *Phys. Rev.*, 1933, **44**, 610.
[6] H. W. B. Skinner, *Proc. Roy. Soc.*, 1933, **A140**, 277.

surfaces in vacuum by evaporation and was also able to achieve reasonably low pressures in his measuring system. His spectra, which were differentiated by hand, quite closely resemble recently measured APS spectra for lithium[7] and beryllium[7,8] with good agreement for the threshold energies. Following the construction of X-ray monochromators, the critical potential method fell into disuse until the 1950's, when it was used again by Shinoda et al.[9] These authors ramped the accelerating potential with a saw-tooth wave and, using simple electronic differentiation, were able to observe appearance potential features with oscilloscope display. Liefeld[10] resurrected the method again in 1967, using more sophisticated electronic differentiation. It was, however, a paper by Park, Houston, and Schreiner in 1970[1] which created fresh interest in the subject and laid the basis for the fruitful work that has followed. The contribution of the latter group was not only to rediscover the method but also to emphasize its surface sensitivity. The application of u.h.v. and clean surface techniques together with electronic differentiation using potential modulation provided a completely new tool in surface studies.

It is worthwhile at this point to mention other appearance potential spectroscopies which have been, or can be applied to surface analysis. Park and Houston have also developed, for example, two forms of Auger electron appearance spectroscopy, both capable of being realized with relatively simple apparatus. In the first form, the threshold energies for the appearance of characteristic Auger electrons are measured in an analogous fashion to the soft X-ray method.[11] The construction of the spectrometer is somewhat more complicated, but its advantage lies in increased sensitivity. (The Auger yield is much greater for all core levels in elements of low atomic number.) Unfortunately at lower energies it would appear that diffraction effects play a role, giving rise to a very uneven background. In the second method[12] the total photoelectron current from the sample is measured as a function of the energy of incident soft X-rays. As in APS, core electron binding energies can be determined from the peaks in the derivative of this curve. The incident photons have a relatively high penetration depth and the measured photoelectrons at the threshold energies consist mainly of secondaries from Auger emission deeper down in the sample. Thus the method is much less surface sensitive than SXAPS, although it is somewhat difficult to assess these parameters quantitively at this stage. A further variation on the experiment has been described by Kirschner and Staib.[13] In their version, called disappearance potential spectroscopy (DAPS), the elastically back-scattered electron current is scanned as a function of energy.

[7] A. M. Bradshaw and W. Wyrobisch, to be published.
[8] P. O. Nilsson and J. Kanski, *Surface Sci.*, 1973, **37**, 700.
[9] G. Shinoda, T. Suzuki, and S. Kato, *Phys. Rev.*, 1954, **95**, 840; *Jap. J. Phys.*, 1956, **11**, 657.
[10] R. J. Liefeld, *Bull. Amer. Phys. Soc.*, 1967, **12**, 562.
[11] J. E. Houston and R. L. Park, *Phys. Rev. (B)*, 1972, **5**, 3808.
[12] J. E. Houston, R. L. Park, and G. E. Laramore, *Phys. Rev. Letters* 1973, **30**, 846.
[13] J. Kirschner and P. Staib, *Phys. Letters (A)*, 1973, **42**, 335.

At the threshold energy for core hole excitation, a sharp *decrease* in the reflection coefficient is expected. The resulting features are qualitatively similar to those in SXAPS or Auger APS. This method should be described as dispersive, however, since one must distinguish between elastically and inelastically back-scattered electrons with an electron energy analyser.

Density of states information similar to that yielded by APS can also be provided by several other methods. *X*-Ray absorption spectroscopy is perhaps the best known and is covered in a recent review by Fabian *et al.*[14] One should note that the method also measures core level binding energies as well as the density of empty conduction band states. Because of the long mean free paths of the incident photons the technique may fairly be described as bulk sensitive. Less well known are the isochromat spectroscopies. Characteristic isochromat spectroscopy, or the measurement of excitation curves, is the dispersive counterpart of the simple SXAPS experiment. Instead of measuring the total photon yield, an *X*-ray spectrometer is used to detect one of the characteristic lines corresponding to the core level.[15-18]

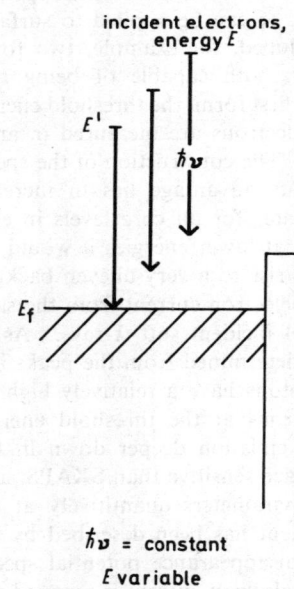

Figure 2 *Principle of the method of continuum isochromat spectroscopy*

[14] D. J. Fabian, L. M. Watson, and C. A. W. Marshall, *Rep. Progr. Phys.* 1972, **34**, 601.
[15] R. J. Liefeld, in 'Soft *X*-Ray Band Spectra', ed. D. J. Fabian, Academic Press, London, 1968, p. 133.
[16] A. F. Burr, *Adv. X-Ray Analysis*, 1970, **13**, 426.
[17] B. Dev and H. Brinkman, *Ned. Tijdschr. Vacuümtechniek*, 1970, **8**, 176.
[18] A. F. Burr, Proceedings of the Conference on Inner Shell Ionisation Phenomena and Future Applications, USAEC, 1973, p. 720.

The advantage lies in the removal of the Bremsstrahlung background making electronic differentiation techniques unnecessary. On the other hand, the signal obtained is rather weak on account of the small acceptance angle of the monochromator. Continuum isochromat spectroscopy[19,20] also provides a method of probing the density of empty states and gives spectra which are relatively easy to interpret: no self-convolution function needs to be unfolded. The principle behind the method is illustrated in Figure 2. The primary energy, E, of the incident electrons is varied as in APS by changing the accelerating voltage between sample and filament. An X-ray spectrometer is used to select a transition of fixed energy $\hbar\nu$, corresponding to a fixed quantity of electron energy being given up as Bremsstrahlung radiation. When $E = E' = \hbar\nu$, the final state is represented by a single slow electron at the Fermi level. Some of the more recent data in this field are compared with APS results later in this chapter. All three last-mentioned techniques share the disadvantage of requiring an X-ray monochromator in the measuring system. Compatibility with u.h.v. equipment becomes difficult and the experiment is much more costly.

2 Underlying Principles of SXAPS

Background Effects.—In the simple yield curve corresponding to the soft X-ray appearance potential spectrum of an element we expect to find step-like features superimposed on the Bremsstrahlung background. When electrons strike a metal target, it is found experimentally that the total integrated Bremsstrahlung intensity at all frequencies is given by [21]

$$I = i_e b Z V (V + 16.3Z) \qquad (1)$$

where Z is the atomic number, V the accelerating potential, i_e the incident electron current (b is a constant, ca. 10^{-9}). Thus in the APS experiments the simple yield curve is expected to rise essentially with the square of the energy of the incident electrons, as is found in practice (for example ref. 22). At the threshold potential for core hole excitation a certain fraction of the incident electrons is used in ionization. These electrons give up all their energy and can therefore no longer produce Bremsstrahlung radiation in the sample. Thus, simultaneous with the sharp increase in characteristic radiation, is a decrease in the accompanying Bremsstrahlung. That the increase in characteristic radiation is the dominant effect follows from the experimental observation of a *positive* step in the simple yield curve. The degree to which we can ignore the negative step in the background will depend on the fluorescence yield for the particular sub-shell and on the

---

[19] K. Ulmer, 'Isochromat Spectroscopy of Transition Metal Alloys', 1972, unpublished.
[20] R. R. Turtle and R. J. Liefeld, *Phys. Rev. (B)*, 1973, **7**, 3411.
[21] E. U. Condon, in 'Handbook of Physics', McGraw-Hill, New York, 1968, p. 7–126.
[22] S. Yamamoto, *Jap. J. Appl. Phys.*, 1973, **12**, 463.

spectral response of the photocathode. In the analysis of the intensity of APS features in the following section, however, the effect will be neglected.

On differentiating equation (1) with respect to V we obtain

$$dI/dV = 2i_e bZV \qquad (2)$$

(The second term can be neglected). Thus in the experimental APS spectrum we expect to observe the peak-like features superimposed on a linearly rising background. In fact the core hole excitation rate will continue to increase at energies higher than the threshold, so the background will contain certain discontinuities, particularly at very high energies, arising from the characteristic radiation.

The Intensity of APS Features.—The intensity relative to background of an APS feature in the simple yield curve as a function of energy is given by the core hole excitation rate, $P(E_0)$ (where $E_0 = eV$, with e the electronic charge) modified by the appropriate value of the fluorescence yield for the particular core level. As explained above, the de-excitation process will not affect the shape or form of the APS feature, merely its absolute intensity. However, it is worthwhile at this point to consider the fluorescence yield and its ramifications in SXAPS. The fluorescence yield may be defined as the probability that a vacancy in an atomic shell is filled in a radiative electron transition from another principal shell.[23] What is relevant in APS is the sub-shell fluorescence yield, ω_i^X, characteristic of a state having a hole in the X_i sub-shell. These quantities are extremely difficult to determine experimentally because the distribution of holes amongst the sub-shells must be known as well as the correct assignment of the characteristic sub-shell X-rays. Thus there are only a few experimental determinations of ω_i^L from $Z = 54$ (Xe) upwards. In elements of low atomic number large differences in ω_i^X for the same shell X can arise owing to selection rules for the change of spin–orbit quantum numbers. The competing de-excitation process is Auger electron emission. The corresponding Auger yield, a, is then defined through $\omega + a = 1$. For all elemental core levels up to several thousand electron volts binding energy the dominant de-excitation mechanism is the Auger process. This can be seen from the value of the average fluorescence yields, $\bar{\omega}_K$, $\bar{\omega}_L$, and $\bar{\omega}_M$, shown as a function of atomic number in Figure 3. For elements of relatively high atomic number, only the higher M or N levels are readily accessible to APS, and here the fluorescence yield is very small indeed. This has not prevented the spectra of thorium[24] and uranium,[25] for example, from being investigated in detail.

To return to the core hole excitation rate, we note that this function is itself dependent on the excitation cross-section, $\sigma(E_0)$, on the spatial dis-

---

[23] W. Bambynek, B. Craseman, R. W. Fink, H. U. Freund, H. Mark, C. D. Swift, R. E. Price, and P. V. Rao, *Rev. Mod. Phys.*, 1972, **44**, 716.
[24] P. A. Redhead and G. W. Richardson, *J. Appl. Phys.*, 1972, **43**, 2970.
[25] R. L. Park and J. E. Houston, *Phys. Rev. (A)*, 1973, **7**, 1447.

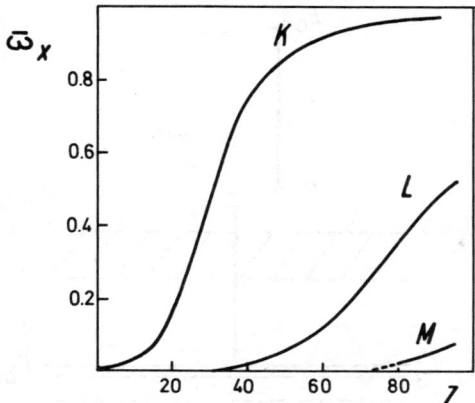

Figure 3 Average fluorescence yields $\bar{\omega}_K$, $\bar{\omega}_L$, and $\bar{\omega}_M$ as a function of atomic number. (Plotted using experimental data tabulated in ref. 23)

tribution in the solid of the atoms of the element in question, and on the penetration depth of the incident electrons. (The analysis that follows is essentially that of Dev and Brinkman[17] and Houston and Park [26, 27].) If we assume that the structure at and near the threshold energy is due entirely to electrons which have created a core hole and been involved in no other loss processes, then we can simply take the elastic mean free path, λ, as the relevant depth parameter. Likewise assuming a uniform distribution, n, of the atoms of the element under investigation we obtain

$$P(E_0) = n\lambda\sigma(E_0) \qquad (3)$$

In writing out the function for the excitation cross-section we are limited at this stage to describing the ground state, Φ, and excited state $f(E_0,E)$ with one-electron wavefunctions,

$$\sigma(E_0) = \int_0^{E_0} \Phi(E - E_b)f(E_0,E)\,\mathrm{d}E \qquad (4)$$

where $\Phi(E - E_b)$ represents the core level function and E_b the core level binding energy (see Figure 4). The function representing the excited state must take into account *both* the electron promoted to the Fermi level *and* the scattered incident electron, and is therefore given by

$$f(E_0,E) = \int_0^{E_0 - E} p_1(\varepsilon_1,E_0)p_2(\varepsilon_2,E_0)N(\varepsilon_1)N(\varepsilon_2)\,\mathrm{d}\varepsilon_2 \qquad (5)$$

where $\varepsilon_1 + \varepsilon_2 = E_0 - E_b$

[26] J. E. Houston and R. L. Park, *J. Chem. Phys.*, 1971, **55**, 460.
[27] R. L. Park and J. E. Houston, *J. Vac. Sci. Technol.*, 1974, in press.

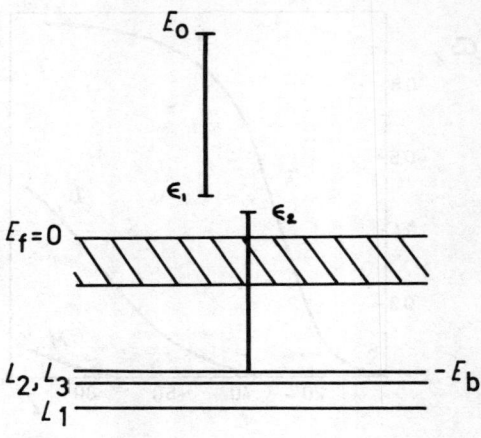

Figure 4 *Energy level diagram for the soft X-ray appearance potential experiment showing symbols used in the text* ($E_0 > E_b$)

$p_1(\varepsilon_1, E_0)$ and $p_2(\varepsilon_2, E_0)$ are the respective transition probabilities and depend as usual on the selection rules. If we can assume that $p_1(\varepsilon_1, E_0)N(\varepsilon_1)$ and $p_2(\varepsilon_2, E_0)N(\varepsilon_2)$ are independent of ε and E_0 at and near the threshold, then they can be taken as simply proportional to $N(\varepsilon)$. Equation (5) reduces to

$$f(E_0, E) = \int_0^{E_0 - E} N(E_0 - E_b - \varepsilon_2) N(\varepsilon_2) \, d\varepsilon_2 \qquad (6)$$

which is simply the self-convolution of the density of empty states above the Fermi level. The experimentally determined features in the simple yield curve would be expected to reflect this function quite closely. Kanski and Nilsson[28] have recently suggested that the neglect of the transition probabilities in equation (5) may seriously oversimplify the physical process. They have compared closely the structure in the derivative of the yield curve for the two levels of barium, N_3 (178.1 eV) and M_5 (782.0 eV). Whilst the overall shapes of the two spectra are similar, there are large differences in the relative amplitude and half-widths of the different features.

[28] J. Kanski and P. O. Nilsson, *Phys. Letters*, in press.

Normally the derivative of the simple yield curve is determined experimentally and so it is useful to differentiate equation (4), taking the core level density of states as a delta function to simplify the expression:

$$\sigma'(E_0) = f'(E_0) = N(0)N(E_0 - E_b) + \int_0^{E_0 - E_b} N(\varepsilon_2) \frac{dN(E_0 - E_b - \varepsilon_2)}{dE_0} d\varepsilon_2 \quad (7)$$

To self-convolute a discontinuous function of this sort we must begin just below the Fermi energy, in which case the first term falls away since $N(0) = 0$. For a simple step-like density of empty states the derivative $dN(E_0 - E_b - \varepsilon_2)/dE_0$ can be approximated by $N_{E_f}\delta E_0$, where N_{E_f} is the density of states at the threshold and δE_0 is a Dirac delta function. Thus

$$\sigma'(E_0) = N_{E_f} \cdot N(E_0 - E_b) \quad (8)$$

The intensity at the threshold is then given simply by,

$$\sigma'(E_0) = N_{E_f}^2 \quad (9)$$

That the approximations involved here are reasonable is illustrated for a simple rectangular function in Figure 5. The peak maximum and its half-width are exactly reproduced in the derivative of the self-convoluted function.

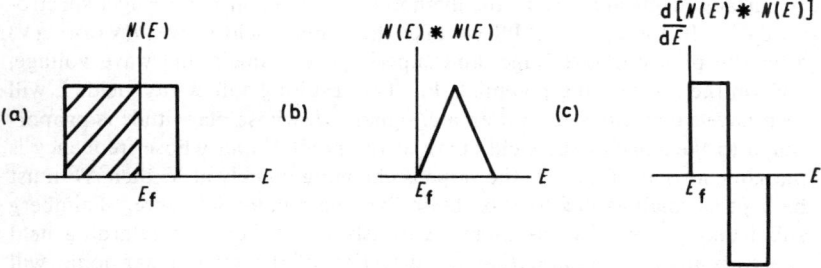

Figure 5 *Self-convolution and differentiation of a simple rectangular density of states function*

Thus we would expect narrow-band metals to give sharp features in APS, containing accessible density of states information. In cases where the structure above the Fermi level is more complicated, the second term can neither be neglected nor assumed to be a constant. It is questionable, therefore, whether the direct comparison by Nilsson and Kanski[29] of complicated APS derivative spectra with band structure calculations is very meaningful.

[29] P. O. Nilsson and J. Kanski, *Electrón Fís. Apli.*, 1974, in press.

In practice, of course, further factors such as lack of monochromaticity of the incident electrons, the modulation voltage, and the presence of discrete loss peaks will further reduce the ease of obtaining useful information. Plasmon loss structure has already been observed for several elements, but this appears to be due to coupling in the final state, rather than with the incident electron. This is discussed more fully below; we should, however, note that this evidence of multielectron behaviour implies that a description in terms of one-electron wavefunctions falls some way short of physical reality. It must also be remembered that the APS experiment may not be probing a 'true' density of states. As Nilsson and Kanski[29] point out, the localized core electron overlaps only part of the conduction states. Thus $N(\varepsilon_2)$, and perhaps also $N(E_0 - E_b - \varepsilon_2)$, in equation (6) would represent only a 'local' density of states.

3 Some Experimental Considerations

Apparatus.—The steps in the SXAPS simple yield curve are quite small in comparison with the Bremsstrahlung background (typically 1%) and difficult to determine in the presence of the accompanying noise. A voltage modulation technique for electronic differentiation offers a way of solving this problem. Although the idea itself is not new, it appears to have been first applied in conjunction with phase sensitive detection by Leder and Simpson.[30] In surface chemical analysis, the method is also familiar from Auger spectroscopy.[31] In the case of APS we take the simple yield curve, $I(V)$ or $i_c(V)$ after the photocathode stage, and superimpose a small sine wave voltage, ΔV, on the accelerating potential, V. The resulting soft X-ray yield, I, will then have superimposed on it an a.c. signal, ΔI, whose magnitude is proportional to the *slope* of the yield curve at the point V and whose frequency is the same as that of ΔV. If the slope is changing quickly at V, then ΔV must be kept as small as possible to obtain the true differential curve. Palmberg and Rhodin[31] discuss this more rigorously for the case of retarding field analysers and further show that the detection of the second harmonic will yield the second derivative of $I(V)$.

A simple form[32] of the APS experiment using this technique is shown in Figure 6 and is essentially similar to that described by Park and Houston.[33] The electron source is a resistively heated tungsten filament situated a few millimetres away from the surface of the sample. The accelerating potential is applied directly between filament and sample and can be scanned from 0 to 1000 V. The soft X-rays emitted from the sample are converted into secondary electrons at the chamber wall and are attracted to the positively

---

[30] L. B. Leder and J. A. Simpson, *Rev. Sci. Instr.*, 1958, **29**, 571.
[31] P. W. Palmberg and T. N. Rhodin, *J. Appl. Phys.*, 1968, **39**, 2425.
[32] A. M. Bradshaw and D. Menzel, *Phys. Status Solidi (B)*, 1973, **56**, 135.
[33] R. L. Park and J. E. Houston, *Surface Sci.*, 1971, **26**, 664.

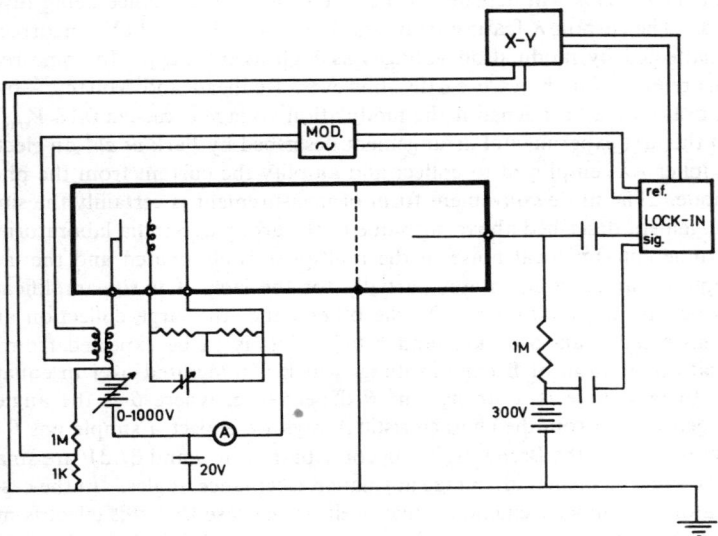

Figure 6 *Measuring circuit for soft X-ray appearance potential spectroscopy using the simple collector mode of operation*

charged collector. The experiment thus derives much of its simplicity from the use of the wall of a vacuum chamber as a photocathode with a large acceptance angle (approx. 135°). Electronic differentiation is performed as described above by applying a small modulation voltage to the accelerating potential and, with a lock-in amplifier, detecting only that part of the signal which corresponds in both frequency and phase. The experimental factors which affect the peak widths in the resulting spectrum are the size of the modulation voltage and the non-monochromaticity of the incident electrons. As far as the latter is concerned, thermal broadening is unavoidable in such an experimental arrangement (typically 0.3 eV). However, the effects of the voltage drop across the filament can be minimized. For instance, the middle section of the filament can be deliberately etched thin to concentrate emission over a small area. With such an arrangement it is possible, using 0.1 mm tungsten wire, to reduce the voltage drop to 0.5 V across the emitting part. Redhead and Richardson[24] used a thoria-coated tungsten emitter in their measurements on thorium surfaces. The low work function gives excellent emission properties at low applied voltages. At the same time heating effects due to radiation at the sample surface are minimized. However, the danger of evaporation of thorium on to the sample exists. In the case of a thorium sample this point is hardly critical. The optimum modulation voltage must be chosen such that the experimentally recorded peaks are seen not to be broadened. This will depend on the levels and the substance being investigated. The carbon K feature from graphite (half-width 2.2 eV, uncorrected) is unaffected by modulation voltages as high as 0.5 V_{rms}. In some recent measurements on the rare earths, however, Redhead and Murthy[34] found that peaks were broadened if the modulation voltage exceeded 0.18 V_{rms}.

In the first experimental arrangement described by Park et al.[1] an electron multiplier was employed to collect and amplify the current from the photocathode. The more convenient form of measurement is certainly the simple arrangement described above, also due to the group at Sandia laboratories.[33] The inherent statistical noise in the multiplier is eliminated and the larger acceptance angle compensates partially for the lack of in situ amplification provided by the multiplier. On the other hand, the large collection angle decreases the peak-to-background ratio. This is to be expected from the spatial distribution of Bremsstrahlung, which is integrated over in equation (1). In fact there is a strong $\sin^2 \theta$ dependence, where θ is the angle of emergence.[35] From the characteristic X-rays we expect a simple $\cos \theta$ distribution. Thus the Bremsstrahlung contributions to I and dI/dV are smaller for the same element and voltage at smaller acceptance angles. In the case of low atomic number elements, it may well be the case that this effect is more important for signal-to-noise (S/N) considerations than the loss of multiplier noise. At present there exist no reliable data to confirm this.

Several modifications to the simple form of the spectrometer have been reported in the literature. For example, a 'compact' appearance potential spectrometer has been described by Musket and Taatjes[36] whereby all the essential features—filament, photocathode, and collector—are mounted on a $2\frac{3}{4}$ inch u.h.v. flange. This has found particular application in the Reporter's laboratory, as it enables the sample to be moved away easily from the measuring system for argon ion bombardment cleaning or for other surface investigations. Long and Beavis[37] report measurements with a spectrometer in which the electron current leaving the photocathode is detected, so that the detector circuit remains near earth. A positively-biased grid in front of the photocathode collects the photoelectrons and at the same time prevents ions reaching the photocathode, thus allowing spectra to be measured at pressures greater than 10^{-7} Torr. Redhead and Murthy[34] have also used a second grid to stop ions reaching the photocathode: operation of this grid at a high potential improved the sensitivity of the method by a factor of at least three compared with operation at 50 V. The modification of a standard LEED–Auger hemispherical grid system to serve as a spectrometer for APS has also been described by Haas et al.[38]

We should at this point, without going into great detail, consider the

---

[34] P. A. Redhead and M. S. Murthy, to be published.
[35] S. T. Stephenson, *Handbuch der Physik*, 1957, **30**, 337.
[36] R. G. Musket and S. W. Taatjes, *J. Vac. Sci. Technol.*, 1972, **9**, 1041.
[37] R. L. Long and L. C. Beavis, *Rev. Sci. Instr.*, 1972, **43**, 939.
[38] T. W. Haas, S. Thomas and G. J. Dooley, *Surface Sci.*, 1971, **28**, 645.

problem of noise as it arises in the experiment. Most of the collected photocurrent at any particular energy consists of Bremsstrahlung, which contributes no useful information but considerable shot noise. A second source of noise arises in the measuring circuit, from the load resistor (1 MΩ in Figure 6). This thermal noise (Johnson noise) can be reduced relatively by increasing the value of the load resistor. The shot noise, on the other hand, is inherent in the signal and is thus more difficult to combat. As the signal is directly proportional to the collected current i_c, but the shot noise to $\sqrt{i_c}$, then the S/N ratio can be improved simply by increasing i_c. This can be done either by increasing the emission current i_e, or, at danger to the resolution, by increasing the amplitude of the modulation voltage. An alternative approach to improving the S/N ratio would be to remove some of the accompanying Bremsstrahlung radiation by employing an energy selective detector. This idea taken to its logical conclusion leads to the measurement of the characteristic isochromat[15-18] described in Section 1, where an X-ray monochromator selects a single, appropriate characteristic emission line. A start in this direction has been made by Verhoeven and Rieger,[39] who used a proportional counter with a Mylar window to obtain a certain degree of energy selection. That the expected major improvements in the S/N ratio are not obtained in their spectra is probably due to the necessarily smaller acceptance angle.

The best photocathode material to use in an appearance potential spectrometer would naturally be the one with the highest photoelectric yield in the soft X-ray region. From the point of view of the photoionization cross-section, it would seem not unreasonable to take a heavy metal such as gold. Using a KCl photocathode, Sims and Foster also report substantial improvements in S/N ratio.[40] Tracy,[41] in a critical appraisal of SXAPS, has discussed the effect of photocathode material on recorded spectra. He noted that a titanium photocathode would produce weak titanium L_3 and L_2 lines in the spectra of other elements. This was found to be due to the short wavelength cut-off of the Bremsstrahlung continuum being modulated back and forth across the titanium L_3 and L_2 ionization thresholds. This effect is likely to be particularly prominent in samples where the short wavelength limit shows sharply defined structure. Exactly the same phenomenon is used to advantage by Houston, Park, and Laramore[12] in their photon-induced APS experiment, which is described in Section 4.

Determination of the Threshold Energy.—One disadvantage of APS lies in the inability to determine *exactly* the core level binding energy. We have already seen that the typical APS derivative spectrum will give us peaks corresponding to the core levels. Of vital importance for the determination of the electron binding energy is the exact threshold energy, or onset of the feature in the simple yield curve. It is readily seen that the effect of the

[39] J. Verhoeven and E. Rieger, *Ned. Tijdschr. Vakuümtechniek*, 1972, **10**, 80.
[40] M. L. Sims and A. I. Foster, private communication.
[41] J. C. Tracy, *J. Appl. Phys.*, 1972, **43**, 4164.

self-convolution integral and subsequent differentiation as well as the width of the core level function can lead to a degree of uncertainty in the interpretation of the experimental results. Indeed, depending on the form of the density of states function, the threshold could even lie nearer to the peak maximum than to the apparent onset in the derivative spectrum. Bradshaw and Menzel[32] have proposed the empirical procedure of taking the intercept of the tangent at half maximum with the extrapolated base line as the threshold value. (An almost opposite view is adopted by Burr[18] in a recent paper discussing exactly this problem as it relates to the measurement of characteristic isochromats. He favours the point of inflection in the isochromat as the threshold point which would thus correspond to the peak maximum in the SXAPS derivative spectrum.) Park and Houston favour the first peak in the second derivative, which of course corresponds to the point of inflection in the first derivative. A further correction for the work function of the filament, ϕ, must be made, since the true incident electron energy exceeds the measured value by $e\phi$. In addition we have the effect of thermal broadening, modulation voltage, and voltage drop across the filament, all of which broaden the features and tend to lower the measured binding energy. The root mean square of the halves of these quantities, as measured in a typical experiment in the Reporter's laboratory, gives 0.45 eV. Taking the generally accepted value of 4.52 V for the work function of tungsten, we would, thus following Park and Houston,[42] favour adding 5.0 eV to the measured threshold energy to obtain the core level binding energy relative to E_f.

4 Applications

Surface Chemical Analysis.—A typical APS spectrum from a dirty titanium surface is shown in Figure 7. Besides the titanium L_3, L_2, and L_1 levels we see carbon, oxygen, nitrogen, and chromium impurities on the surface. Tracy[43] has compared a similar titanium spectrum with the corresponding Auger spectrum from the same sample and, amongst other differences in relative intensity, was unable to detect sulphur in the appearance potential spectrum. (The whole spectrum is not shown in Figure 7, but frequent attempts under conditions of high sensitivity to observe sulphur peaks on this surface were also unsuccessful). The frequent insensitivity to sulphur, a normal impurity on metal surfaces, appears to be one of the drawbacks to APS in surface analysis. In a recent comment on Tracy's papers, Park and Houston[44] publish a spectrum from a niobium surface showing a very strong sulphur peak; similarly, Redhead and Richardson's thorium spectra[24] indicate sulphur quite clearly, as is shown in Figure 8. The reason for this anomalous behaviour is not yet clear. It was also previously thought that heavy metals and semiconductors were particularly insensitive in APS: high

---
[42] R. L. Park and J. E. Houston, *Phys. Rev.* (*B*), 1972, **6**, 1073.
[43] J. C. Tracy, *Appl. Phys. Letters*, 1971, **19**, 353.
[44] R. L. Park and J. E. Houston, *J. Appl. Phys.*, 1973, **44**, 3810.

Appearance Potential Spectroscopy and Related Techniques

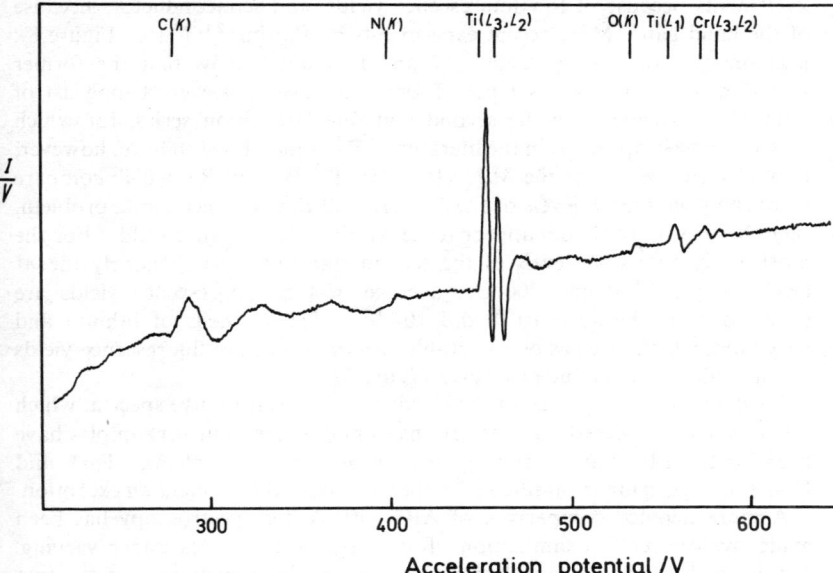

Figure 7 *Soft X-ray appearance potential spectrum of a dirty titanium surface in the range 200—700 eV*

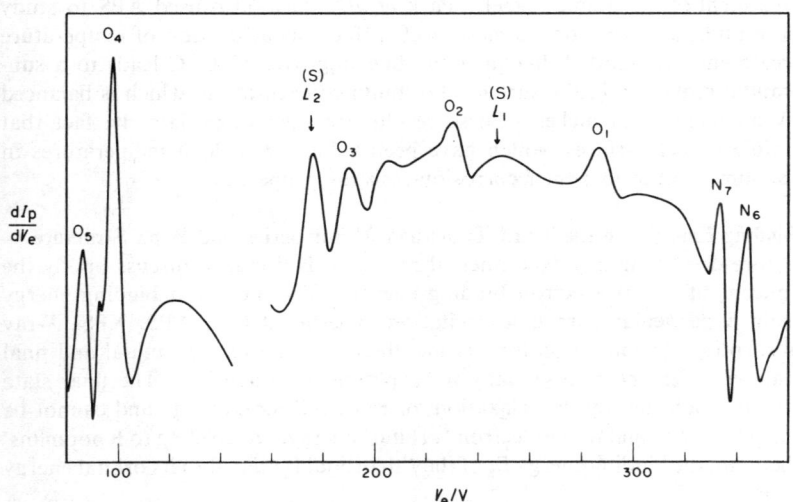

Figure 8 *Soft X-ray appearance potential spectrum of thorium on tungsten in the energy range 80—360 eV. The portion below 150 eV was taken with the sensitivity reduced by 12.5*
(Reproduced by permission from *J. Appl. Phys.*, 1972, **43**, 2971)

Z elements because of low fluorescence yields, and semiconductors because of the band gap. More recent experiments on thorium[24, 34] (see Figure 8), uranium,[25] certain rare earths[34, 45] and niobium[44] show that the former assertion is not necessarily true. There still exists, however, a long list of elements, particularly in the second and third transition series, for which no spectra have appeared in the literature. Park and Houston have, however, unpublished spectra for the M levels of Hf, Ta, W, and Re, which compare favourably with the L levels of the $3d$ series. If this is indeed a real problem, then the answer would not appear to lie with the fluorescence yield. For the most easily accessible levels of the second transition series, namely the M levels between 200 and 700 eV, average M-shell fluorescence yields are expected to lie between 10^{-4} and 10^{-3}.[23] The K shells of lithium and beryllium, which give easily observable spectra, also have fluorescence yields of this order of magnitude (see also Figure 3).

Insulating or semiconducting oxide layers *do* appear to give spectra, which suggests that unsuccessful investigations of bulk semiconductor samples have been hindered by poor sensitivity, or even by contact problems. Park and Houston's spectrum from silicon[44] appears to be at the moment an exception.

A more detailed comparison of APS with Auger spectroscopy has been made by Musket.[46] Examination of stainless steel surfaces under varying conditions leads to the conclusion that the relative sensitivities of the two techniques are neither constant nor absolute. The sensitivity ratios for particular impurity atom levels vary considerably according to the pretreatment of the stainless steel. Park *et al.*[47] have also used APS to study chromium depletion on stainless steel surfaces as a function of temperature treatment. It is found that vacuum annealing over 1000 °C leads to a substantial reduction in the surface chromium concentration, which is balanced by an increase in nickel. These results are used to explain the fact that stainless steel surfaces, which have been subjected to high temperatures in vacuum, tend to lose their corrosion-resisting properties.

Binding Energies in the Third Transition Metal Series and Band Structure.— Before describing any experimental results, it is useful to discuss briefly the concept of a core electron binding energy. The measured binding energy from a particular core hole excitation experiment (*e.g.* APS, XPS, X-ray absorption, or ionization loss) is the difference between the initial and final state energies and is essentially an experimental quantity. The final state energy is affected by the relaxation, or reorganization energy and cannot be simply be set equal to the electron 'orbital' energy. According to Koopmans' theorem, the binding energy E_b of the j'th orbital is related to its orbital energy E' by

$$E_b(j) \sim E'(j) \tag{10}$$

[45] M. B. Chamberlain and W. L. Baun, to be published.
[46] R G. Musket, *J. Vac. Sci. Technol.*, 1972, **9**, 603.
[47] R. L. Park, J. E. Houston, and D. G. Schreiner, *J. Vac. Sci. Technol.*, 1972, **9**, 1023.

This statement cannot be exact because relaxation of the outer orbitals towards the hole in the final state is not taken into consideration. In many XPS experiments on gas-phase species the measured binding energies are thus smaller than the best self-consistent field orbital energies.[48] In the solid state it is not known how important relaxation effects are, but in different experiments it is clear that the final states will be quite different. Thus in APS we have a core hole and two slow electrons 'localized' on the atom in the final state; in XPS we have a core hole and a single fast electron which may still be located near the hole depending on its velocity compared with the core hole lifetime. Relaxation effects could therefore account for the general result that APS binding energies are lower than the corresponding XPS binding energies. In addition the absolute XPS values require a precise determination of the work function of the analyser slits. (In the case of the Siegbahn data, on which the Bearden and Burr[49] tables are based, the material was oxidized copper.) Any error in this value could also introduce a systematic difference between the two sets of data. By the same token, an incorrect method to estimate the threshold energy from APS data would introduce a systematic error.

The photon-induced APS experiment of Houston et al.[12] may also provide an example of a case where final state effects cannot be neglected. These authors accelerate electrons on to a tungsten target with variable voltage, V, and superimpose on it a small modulation voltage. The resulting radiation, mainly Bremsstrahlung, is allowed to fall on to the sample under investigation. When $eV = E_b$, the Bremsstrahlung short wavelength cut-off coincides with the ionization energy and the effect of the modulation is to shift the cut-off back and forth across this excitation threshold. Synchronous detection of the electron current from the sample gives a first derivative photon-induced Auger appearance potential spectrum. The success of the method relies on the shape of the short wavelength cut-off in the chosen converter material, which is, incidentally, also dependent on the density of empty states above the Fermi level. For tungsten this is reasonably sharp,[50] but does appear to contain a certain amount of structure, as can be seen from Figure 9. Comparison of the SXAPS and the photon-induced Auger APS spectra for nickel is shown in Figure 10. A difference in threshold energy of 0.9 eV is quite clearly seen. Houston et al.[12] account for this in terms of a surface versus bulk chemical shift. (The sampling depth in photon-induced Auger APS is much larger than in SXAPS.) Recent XPS measurements on clean tungsten at various angles of incidence[51] show, however, no evidence for surface tungsten atoms having lower electron binding energies. The operation of a relaxation energy effect would provide an explanation: the final state

---

[48] See, for example, W. A. Richards, *Internat. J. Mass Spectrometry Ion Phys.*, 1969, **2**, 419; U. Gelius, G. Johanssen, H. Siegbahn, C. J. Allan, D. A. Allison, and J. Allison, *J. Electron Spectroscopy*, 1973, **1**, 285; D. A. Shirley, *Adv. Chem. Phys.*, 1973, **23**, 1973.
[49] J. A. Bearden and A. F. Burr, *Rev. Mod. Phys.* 1967, **39**, 125.
[50] G. Böhm and K. Ulmer, *J. de Phys. Colloque (C4)*, 1971, **32**, 241.
[51] A. M. Bradshaw, D. Menzel, and M. Steinkilberg, *Jap. J. Phys.*, 1974, in press.

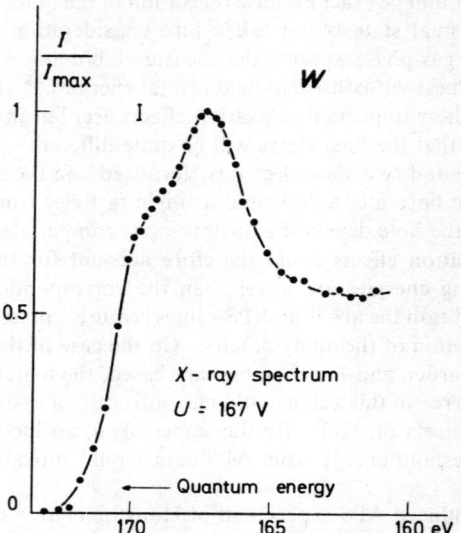

Figure 9 Short wavelength limit of the X-ray continuum spectrum from a tungsten surface. Incident electron energy = 170 eV
[Reproduced by permission from *J. de Phys. Colloque* (C4), 1971, **32**, 241]

in the photon-induced experiment is a core hole with only *one* slow electron near the Fermi level.

Park and Houston[42] have recently published a complete series of L shell appearance potential spectra under good vacuum conditions for the transition metals scandium through to nickel. The measured binding energies for the L_1, L_2, and L_3 sub-shells have been tabulated and compared with the Bearden and Burr values.[49] In every case the APS value lies below that of the corresponding XPS-based value, although the probable errors overlap in some cases. As was shown in Section 2, the width of the first positive peak in the APS derivative spectrum will correspond to the width of the unfilled part of the conduction band, if the latter is a sharply defined, reasonably symmetric function. This condition appears to be fulfilled in the case of the $3d$ transition series. Park and Houston obtain an almost linearly decreasing half-width for the L_3 peak in the series scandium (3.3 eV) through to nickel (0.9 eV, both these values are corrected for the instrument response function). This is in good agreement with the qualitative picture of the filling of a rigid d-band, as one progresses along the transition series. In addition there is reasonable agreement with the expected width taken from the APW band structure calculations of Snow and Waber.[52] At the high-Z end of the series a large contribution to the measured width is expected to come from lifetime

[52] E. C. Snow and J. T. Waber, *Acta Metallurgica*, 1969, **17**, 623.

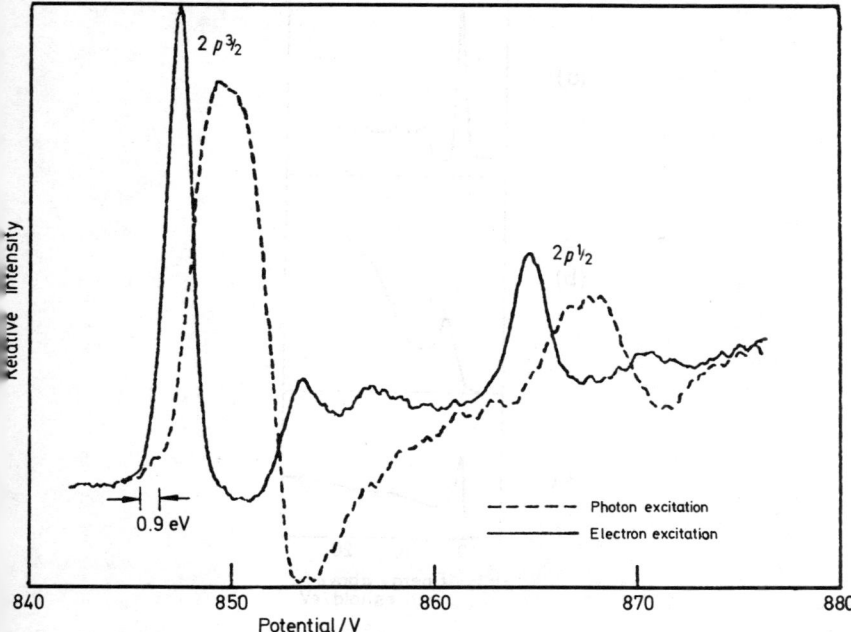

Figure 10 *Comparison of the 2p electron-excited soft X-ray appearance potential spectrum and the photon-induced Auger appearance potential spectrum. The difference in threshold energy for the $2p_{3/2}$ level (L_3) is about 0.9 eV* (Reproduced by permission from *Phys. Rev. Letters*, 1973, **30**, 848)

broadening of the core level function. As an example of how these data compare with those from other techniques for probing the unoccupied density of states, Figure 11 shows a comparison of the APS derivative spectrum for nickel L_3 with the characteristic isochromat for nickel L_3 and the nickel continuum isochromat. Bearing in mind the different quantities being measured, there is excellent correspondence between the APS derivative curve (a) and the continuous isochromat (c). As the continuous isochromat measures $N(E)$ directly, this result is justification enough for the application of APS in band structure studies and for the validity of equations (7) and (8) in Section 2.

Further useful band structure information is obtained from consideration of the $L_3 : L_2$ intensity ratios. Park and Houston's data show clearly that this does not remain constant for the 3d transition series: there is substantial deviation from the expected ratio of 2 based on a statistical weighting of $2j + 1$. This would indicate that there is either a substantial difference in the sub-shell fluorescence yields, ω_3^L and ω_2^L, or that the excitation probability is not governed by the $2j + 1$ weighting. The second possibility has been

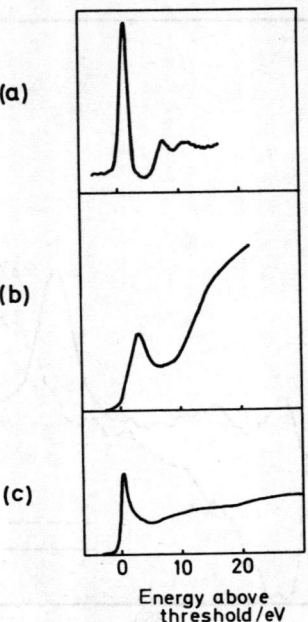

Figure 11 *Comparison of* (a) *the nickel L_3 soft X-ray appearance potential spectrum*[42] *with* (b) *the nickel L_3 characteristic isochromat*[17] *and with* (c) *the nickel continuum isochromat*[20]

eliminated in another experiment of Houston and Park's, where they compare the SXAPS and Auger APS spectra for chromium[11] (see Figure 12). The Auger APS $L_3 : L_2$ ratio is almost exactly two (one can assume that this accurately expresses the relative excitation probabilities because the vast majority of core holes will decay *via* the Auger process), whereas the SXAPS $L_3 : L_2$ ratio is approximately one. The fact that $\omega_3^L \neq \omega_2^L$ implies the presence of j-dependent selection rules for the radiative decay valence band $\rightarrow L$ shell.

Carbon and Plasmon Coupling Phenomena.—Appearance potential spectra from carbon surfaces have been measured by Houston and Park[2,53] and Bradshaw and Menzel.[32,54] A typical spectrum from the basal plane of pyrolitic graphite is shown in Figure 13. The first sharp peak represents the K level excitation. The binding energy is given by 283.6 (± 0.2) eV, compared with the Bearden and Burr value of 283.8 (± 0.4) eV. Any difference here is contained inside the probable experimental error. The main feature of Figure 13 is the remarkably strong satellite structure, indicating a coupling

[53] J. E. Houston and R. L. Park, *Solid State Comm.*, 1972, **10**, 91.
[54] A M Bradshaw and D. Menzel, Proceedings of the International Carbon Conference, Baden-Baden, 1972, p. 101.

Appearance Potential Spectroscopy and Related Techniques 173

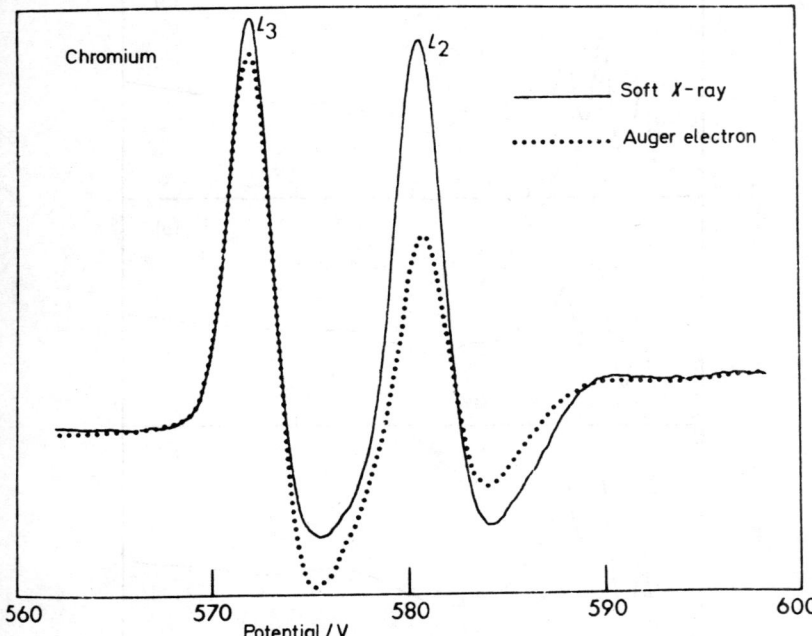

Figure 12 *Comparison of the soft X-ray and Auger appearance potential spectra for the L_3 and L_2 levels from a clean chromium surface*
[Reproduced by permission from *Phys. Rev. (B)*, 1972 **5**, 3808]

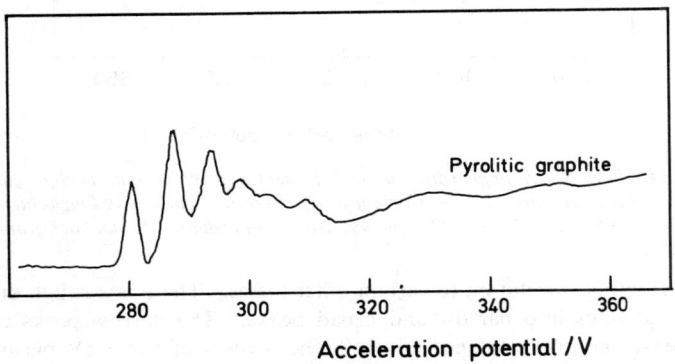

Figure 13 *Soft X-ray appearance potential spectrum from the basal plane of stress-annealed pyrolitic graphite (Union Carbide)*

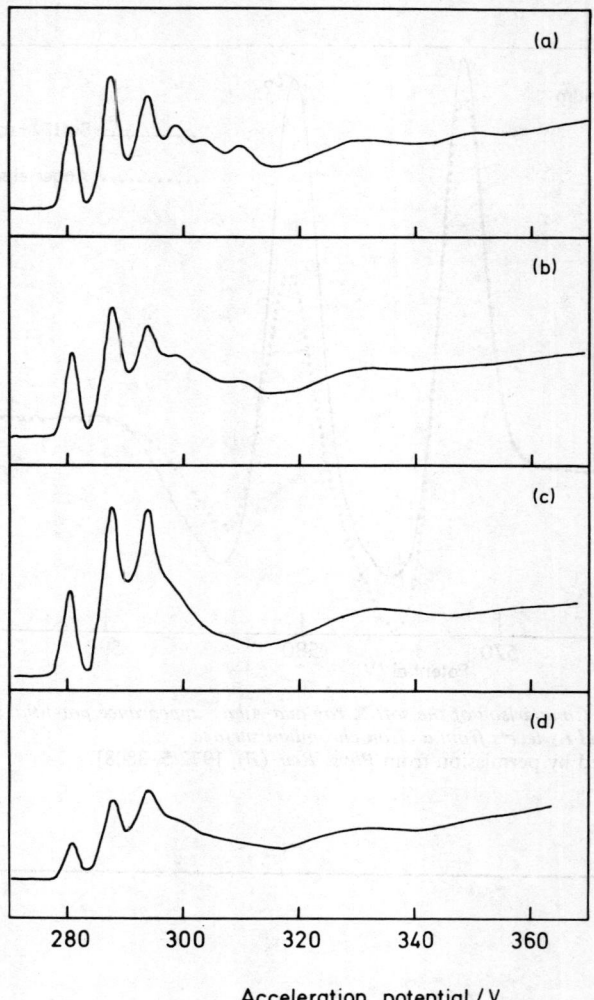

Figure 14 *Soft X-ray appearance potential spectra from various carbon surfaces:* (a) *pyrolitic graphite (Union Carbide)*; (b) *pyrolitic carbon (Glanzkohlenstoff)*; (c) *Sigri L Y4–HTT* 940° C; (d) *vitreous carbon (Sigradur).* Actual backgrounds are not shown

of the core hole excitation to certain other losses. There is a rough division of the satellites into narrow and broad peaks. The narrow peaks can be assigned to coupling to successively higher orders of the 7 eV plasmon,[55] except the peak at 310 eV, which Bradshaw and Menzel attribute to the

[55] K. Zeppenfeld, *Z. Phys.*, 1968, **211**, 391.

27—29 eV plasmon. The underlying hump and the further broad peaks at 330 and 350 eV are attributed to a coupling with interband transitions. The theoretical considerations of Langreth[56, 57] suggest that the high coupling constant results from a strong final state interaction. The known anisotropy of the plasma oscillation in graphite enables one to differentiate between initial and final state coupling effects.[32] Spectra from the prismatic face of pyrolitic graphite, where the direction of momentum transfer in plasmon excitation would be predominantly perpendicular to the c-axis, are identical with spectra from the basal plane, where the direction of momentum transfer would be predominantly parallel to the c-axis. As the plasmon varies both in energy and intensity between these two extremes, it is assumed that coupling with the incident electron does not play a great role. Coupling with the two slow electrons and the core hole (a final state, where slow charge is not conserved during the transition[58]) thus seems to be the primary mechanism. The concept of slow charge conservation appears to be important for the occurrence of a high coupling constant. XPS measurements on pyrolitic graphite show only weak 6.8 and 28 eV plasmon satellites[59] whereas recent measurements of the ionization loss spectrum[60, 61] indicate a very high coupling constant. A further interesting feature of the spectrum in Figure 13 is the fact that the separation between the first five sharp peaks becomes successively, but not regularly, smaller. This has been attributed to anharmonicity or to the consequence of the plasmon anisotropy.

A high degree of graphitic ordering appears to be required in order to obtain a strongly peaked satellite structure. Appearance potential spectra from four different carbons are shown in Figure 14. Houston and Park[53] obtained similar spectra by argon ion bombardment of a graphite surface. Sims and Foster[40] have studied the build-up of carbon in a platinum single crystal surface left overnight in a pressure of $<10^{-9}$ Torr. The carbon K excitation produced a *single* peak of half-width approximately 2.7 eV. Only on annealing at typically 1020 K did the first two plasmon satellites become apparent.

High plasmon coupling constants have also been observed for boron adsorbed on to nickel[2] (see Figure 15) and in nickel oxide.[62] Nagel[63] notes that a better calculated APS nickel L_3 spectrum is obtained when a certain degree of plasmon coupling is included, but in general the effect seems to be restricted to non-metals and to the simple metals, described below.

[56] D. C. Langreth, *Phys. Rev. Letters* 1971, **26**, 1229.
[57] J. J. Chang and D. C. Langreth, *Phys. Rev.* (*B*), 1972, **5**, 3512.
[58] G. E. Laramore, *Solid State Comm.* 1972, **10**, 85.
[59] A. M. Bradshaw and H. Lohneiss, unpublished results.
[60] J. E. Houston and R. L. Park, private communication.
[61] J. Kirschner, private communication.
[62] J. E. Houston and R. L. Park, in 'Electron Spectroscopy', ed. D. A. Shirley, North Holland, Amsterdam, 1972, p. 895.
[63] D. J. Nagel, D. A. Papaconstantopoulos, and J. W. McCaffrey, Proceedings of the International Symposium on X-Ray Spectra and Electronic Structure of Matter, Munich, 1972.

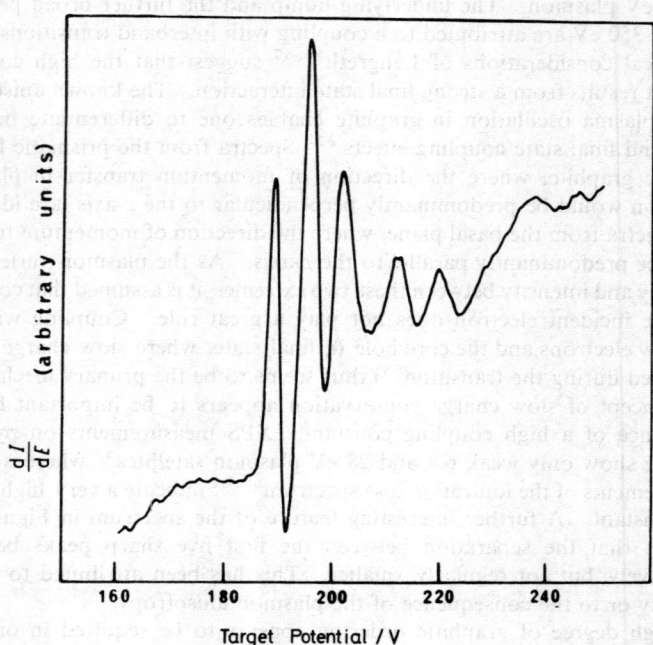

Figure 15 *Soft X-ray appearance potential spectrum for the boron K level from a contaminated nickel surface, showing similarity to the graphite spectrum of Figure 13 (Reproduced by permission from J. Vac. Sci. Technol., 1971, **8**, 91)*

Simple Metals.—Nilsson and Kanski[8, 28, 29, 64] have measured APS spectra for the elements beryllium, magnesium, aluminium, calcium, strontium, and barium. Relatively complicated peak structures above the threshold are observed and the authors tend to favour a density of states explanation in terms of the simple theory. It is, however, doubtful as to whether the direct comparison of complicated APS features with band structure calculations, or with X-ray absorption spectra, is valid. This is discussed briefly in Section 2. Self-convolution and differentiation of relatively simple trial functions show quite quickly that peaks in $N(E)$ do not necessarily occur in the same position, if at all, in $d[N(E) * N(E)]/dE$. An alternative explanation for the beryllium K spectrum[7] attributes the structure to a combination of density of states features and loss peaks. The spectrum, shown in Figure 16, resembles very strongly the plasmon dominated spectra of Figures 13—15. The corrected binding energy is given as 111.7 (± 0.3) eV compared with 111.0 (± 1.0) eV in Bearden and Burr's tables,[49] and 111.7 (± 0.4) eV from

---

[64] P. O. Nilsson and J. Kanski, Abstract 33rd Physical Electronics Conference, Berkeley, 1973.

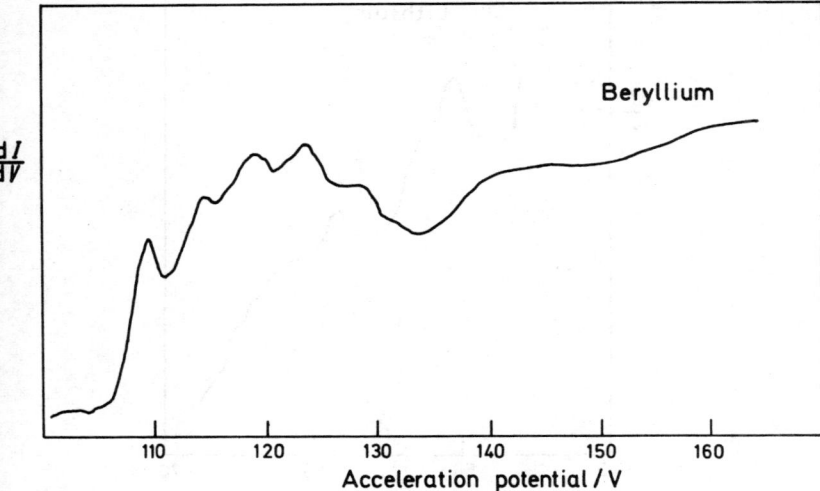

the ionization loss measurements of Swanson and Codling.[65] There is a considerable measure of agreement between the loss structure in Swanson and Codling's paper and that of Figure 16.

Bradshaw and Wyrobisch[7] have also measured a lithium K spectrum (Figure 17) which, surprisingly enough, is almost identical with Skinner's spectrum from the year 1932.[4] The binding energy is determined as 54.3 (± 0.3) eV compared with the Bearden and Burr value of 54.75 (± 0.02) eV.[49] The surface was produced freshly *in vacuo* before each measurement by mechanical scraping. Even then a shoulder 2.6 eV from the main peak remains, which is due to lithium oxide or hydroxide. This peak grows during the course of the measurement as a result of the relatively poor vacuum conditions. (The background pressure was 10^{-8} Torr during the experiment: baking temperatures could not exceed 120 °C on account of the high lithium vapour pressure). The third peak is undoubtedly due to a coupling of the volume plasmon with the core hole excitation and would appear to be the first unambiguous example in APS of a high coupling constant in a metal. The plasmon energy is measured as 7.5 eV, compared with Fellenzer's value of 8 eV from energy loss experiments.[66]

Alloys.—The rigid band model for binary alloys requires that the two components share a common conduction band, and that changing the alloy composition merely varies the extent to which the band is filled. The APS

---

[65] N. Swanson and K. Codling, *J. Opt. Soc. Amer.* 1968, **58**, 1192.
[66] H. Fellenzer, *Z. Phys.* 1961, **165**, 419.

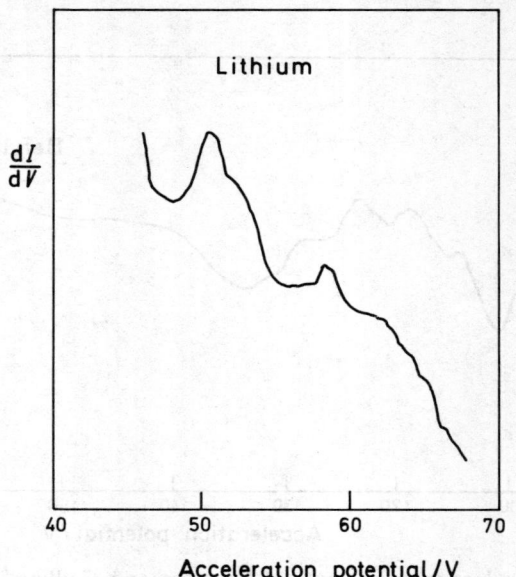

Figure 17 *Soft X-ray appearance potential spectrum for the lithium K level*

investigations of Ertl and Wandelt[67] on the copper—nickel system indicate, in common with several other spectroscopic methods, that this model has considerable shortcomings. The appearance potential spectra in the nickel L_3 and L_2 regions from pure nickel and four different alloy compositions are shown in Figure 18. Although the rigid band model predicts a complete filling of the d-band at high copper concentrations, sharply defined nickel peaks are still to be seen at a 75% copper composition. Furthermore no peak at 931 eV (the expected copper L_3 binding energy) is observed at any of the alloy compositions, just as in the case of the pure metal. Thus despite the presence of an unfilled d-band, it is still not possible to observe features from the excitation of copper core holes. Using equation (9), Ertl and Wandelt plotted N_{E_f} against alloy composition and obtained a good fit to the theoretical curve calculated by Stocks *et al.*[68] using the coherent potential approximation. In this model the constituents retain to a certain extent the electronic configuration of the pure phases.

We should also perhaps make the general reservation that alloy composition in the surface region may not reflect that of the bulk. Of greatest danger to the interpretation of experimental results, such as those of Ertl and Wandelt, would be a two-dimensional separation into the pure phases at the surface.

[67] G. Ertl and K. Wandelt, *Phys. Rev. Letters*, 1972, **29**, 218.
[68] G. M. Stocks, R. W. Williams, and J. S. Faulkner, *Phys. Rev.* (*B*), 1971, **4**, 4390.

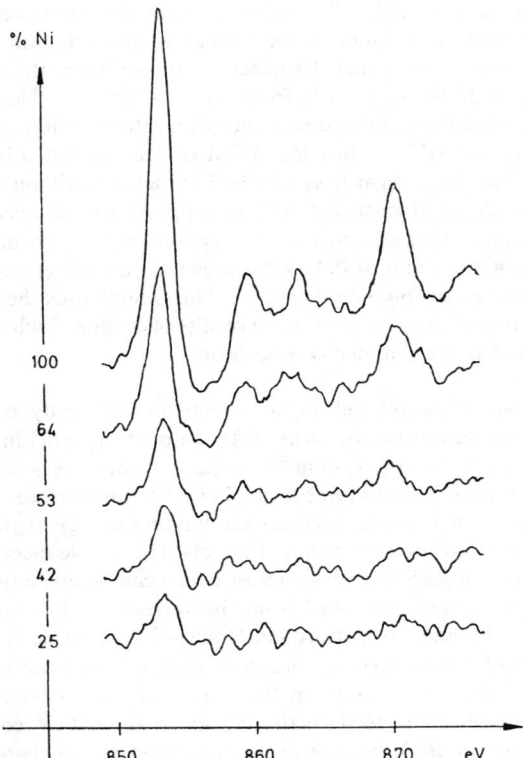

Figure 18 *Soft X-ray appearance potential spectrum for the nickel L_3 and L_2 levels from pure nickel and nickel/copper alloys containing 64, 53, 42, and 25% Ni* (Reproduced by permission from *Phys. Rev. Letters*, 1972, **29**, 218)

It ought to be mentioned again that the APS experiment is probably measuring a local density of states: the final state involves two slow electrons localized at the ionized atom. The degree to which this local density of states approximates to the 'true' density of states is at the moment an unknown factor. This difference could account for the copper–nickel results and not invalidate the rigid band model. On the other hand, recent XPS data[69] do seem to back up the first, simpler explanation of the APS results. Photoemission experiments probe a 'total' density of (filled) states. Houston and Park's data from the titanium–nickel system[70] suggest that APS does in fact sample a total density of states. Their samples were prepared using a surface-alloying technique in which a thin layer of one metal is evaporated

[69] S. Hüfner, G. K. Wertheim, R. L. Cohen, and J. H. Wernick, *Phys. Rev. Letters*, 1972, **28**, 488.
[70] J. E. Houston and R. L. Park, *J. Vac. Sci. Technol.*, 1972, **9**, 579.

on to the other and heated. The nickel L_3 peak for dilute concentrations on a titanium surface is found to be similar to the titanium L_3 peak for dilute concentrations on a nickel surface. Furthermore, the width of the peak is intermediate between those from the pure phases. The peak shapes are most alike when the estimated concentration ratio is unity, corresponding to the stable phase TiNi. Thus the width of the unfilled d-band in TiNi appears to be the same, regardless of whether nickel or titanium core holes are excited, indicating at least that APS is sampling a total density of states. In the alloy composition assumed to correspond to TiNi, chemical shifts of 0.6 eV for titanium L_3 and of 0.8 eV for nickel L_3 are observed, *both* in the direction of increased binding energy. This anomalous behaviour also appears in the recent soft X-ray emission results of Källne,[71] who investigated a TiNi sample of predetermined composition.

Chemical Shifts.—Although relaxation effects should really be taken into account, we will assume that chemical shifts can be interpreted in terms of the initial state properties of the system.[48] Figure 17 shows a good example of a chemical shift occurring in atoms located at a lithium surface. The formation of an oxide or hydroxide increases the binding energy of the K level by 2.6 eV. Siegbahn and co-workers have formulated a simple theory to account for this behaviour in XPS,[72] the essence of which can be taken over and used in APS. When charge is removed from the valence shell of an atom (*e.g.*, when lithium is bonded to a more electronegative atom such as oxygen), the potential field experienced by the core electrons is lowered and their binding energies increase. Thus, in the same way, heavy oxidation of the beryllium surface of Figure 16 shifts the K peak by about 3 eV, corresponding to a similar shift in BeO relative to Be measured by Siegbahn.[72] Other examples of chemical shifts occurring on oxidation are given by Houston and Park[62] in spectra for titanium, chromium, and nickel surfaces. Mild titanium oxidations have also been studied by Kirschner and Staib[73] using the disappearance technique. One of their second derivative $L_3 : L_2$ spectra is shown in Figure 19. It is possible to account for the peak splitting (separation 1.9 eV) in terms of oxidized titanium and unoxidized titanium deeper in the bulk. Alternatively, the nucleation stage of oxidation could have been reached, whereby small islands of oxide cover the surface.

Chemical shifts in the refractory metal compounds have been measured by Bradshaw[74] in APS and, earlier, by Ramqvist[75] in XPS. An appearance potential spectrum for the titanium L_3 and L_2 levels in Ti metal, TiC, and TiN is shown in Figure 20. The corrected binding energies are 453.6 (± 0.3),

---

[71] E. Källne, Uppsala University Institute of Physics Report No. 824, June 1973.
[72] K. Siegbahn *et al.*, 'Electron Spectroscopy for Chemical Analysis', Nova Acta Regiae Societatis Scientorum Upsaliensis, Ser. IV, 1967, Vol. 20.
[73] J. Kirschner and P. Staib, unpublished results.
[74] A. M. Bradshaw, to be published.
[75] L. Ramqvist, K. Hamrin, G. Johanssen, A. Fahlman, and C. Nordling, *J. Phys. Chem. Solids*, 1969, **30**, 1835.

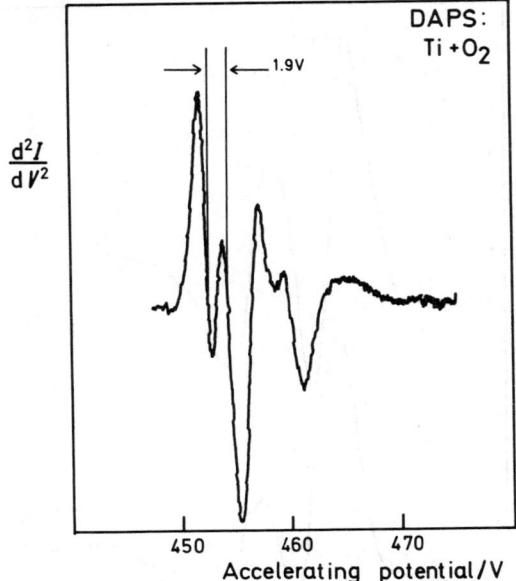

Figure 19 Second derivative 'disappearance' potential spectrum for the titanium L_3 and L_2 levels from a partially oxidized titanium surface

454.8 (±0.3), and 454.8 (±0.3) eV, respectively, indicating chemical shifts towards increasing binding energy in both TiC and TiN of 1.2 eV. Ramqvist reports shifts of 1.3 and 1.5 eV for TiC and TiN, respectively. The nitrogen region in the TiN spectrum consists of three weak features separated by 8.5—9.0 eV; the threshold for the first is located at 397.0 (±0.5) eV, which corresponds closely to the Ramqvist spectrum. The apparent direction of charge transfer, from metal to carbon (nitrogen), is in agreement with the APW band structure calculations of Ern and Switendick,[76] Conklin and Silversmith,[77] and Neckel.[78] The APS results would, however, indicate a high density of states at the Fermi level, and an unoccupied d-band, 3.0—3.5 eV wide.

5 Summary

In this review we have attempted to show that, despite the only very recent rediscovery of the technique, appearance potential spectroscopy has already found several interesting applications in surface studies. First and foremost

---

[76] V. Ern and A. C. Switendick, *Phys. Rev. (A)*, 1965, **137**, 1927.
[77] J. B. Conklin and D. J. Silversmith, *Internat. J. Quant. Chem.*, 1968, **2**, 243.
[78] A. Neckel, P. Rastl, K. Schwarz, and R. Eibler-Mechtler, *Z. Naturforsch.*, 1974, **29a**, 107.

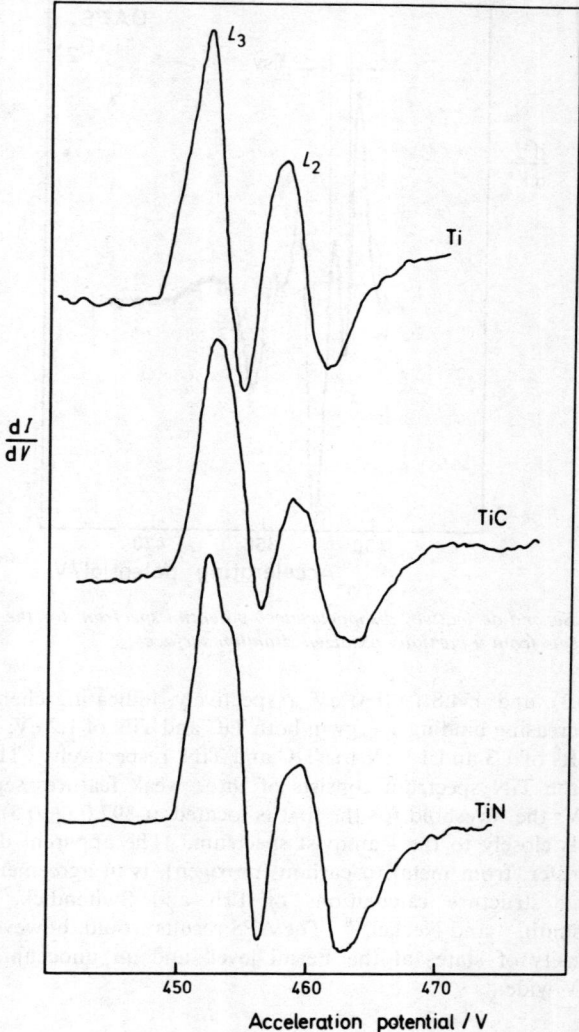

Figure 20 *Soft X-ray appearance potential spectra for the titanium L_3 and L_2 levels from* Ti *metal*, TiC *and* TiN

there now exists a very simple method of obtaining a self-consistent set of core electron binding energies. As in XPS, the application of the technique to surface analysis follows as a logical consequence. The debate as to the sensitivity of APS compared with that of other surface analysis techniques such as XPS and Auger has not yet been resolved. Earlier assertions, not without foundation, that spectra from heavy elements, semiconductors, and insulators were obtainable, now appear not to be entirely correct. Nevertheless, there is still a long list of metals for which spectra have not yet been reported. Secondly, we have a method of determining chemical shifts in electron binding energies in atoms at solid surfaces. A large field of application here in alloys and pseudo-metallic compounds has only just started to open up. Band structure studies provide a third major area of application. The Houston and Park results from the $3d$ transition series have been discussed above in some detail. Further studies of the $L_3 : L_2$ intensity ratios are expected to provide more useful information on the filled part of the conduction band. The refractory metal compounds also seem to provide a fruitful area of band structure investigations with APS.

6
Some Aspects of the Nature and Reactivity of Adsorbed States of Unsaturated Hydrocarbons on Metal Catalysts

BY G. WEBB

1 Introduction

Since the early work by Sabatier et al.[1] on the hydrogenation of ethylene over metal catalysts, the metal-catalysed hydrogenation of unsaturated hydrocarbons has been the subject of a vast number of studies. The metals most commonly used for these studies are the Group VIII metals and copper, in the form of evaporated films, wires, foils, powders, or supported on a carrier such as alumina, silica, or carbon. Recently this range of catalysts has been extended to include rhenium[2] and gold,[3] both of which possess some activity for ethylene hydrogenation.

Many of the early studies were devoted to the establishment of the mechanisms of catalysed reactions with emphasis on the procurement of kinetic data, product distributions, and isotopic distributions obtained from the use of deuterium as a tracer. More recently, however, more attention has been paid to studies of (i) the nature and reactivity of the adsorbed states of unsaturated hydrocarbons, and (ii) the influence of the nature and structure of a catalyst upon its catalytic behaviour. This Report is concerned with these aspects of the interactions of unsaturated hydrocarbons with metal surfaces.

2 Nature of the Adsorbed States of Unsaturated Hydrocarbons

Information regarding the nature of the adsorbed states of hydrocarbons has been obtained from three main sources: (i) indirectly from studies of surface processes such as hydrocarbon–deuterium exchange,[4] self-hydrogenation, and self-poisoning; (ii) from changes in the properties of the metal, e.g. magnetic susceptibility and work function, during and after adsorption; and (iii) by direct observation using i.r. spectroscopy. Most of these studies have been concerned with ethylene and acetylene as adsorbates, although some studies of higher hydrocarbons have also been reported.

It is well established that, when ethylene is adsorbed on to a freshly prepared

---

[1] P. Sabatier and J. B. Senderens, *Comp. rend.*, 1902, **135**, 87.
[2] J. Grant, R. B. Moyes, and P. B. Wells, *J. C. S. Faraday Trans.*, 1973, **69**, 1779.
[3] G. C. Bond, P. A. Sermon, D. A. Buchanan, G. Webb, and P. B. Wells, *J. C. S. Chem. Comm.*, 1973, 444.
[4] G. C. Bond and P. B. Wells, *Adv. Catalysis*, 1964, **15**, 171.

metal surface, self-hydrogenation resulting in the fast production of ethane is observed.[5-8] This formation of ethane may arise either by reaction of adsorbed ethylene with hydrogen atoms released by dissociation of associatively adsorbed ethylene, viz.:

$$C_2H_4\,(a) \longrightarrow C_2H_3\,(a) + H\,(a)$$
$$C_2H_3\,(a) \longrightarrow C_2H_2\,(a) + H\,(a)\ etc.$$
$$C_2H_4\,(a) + 2H\,(a) \longrightarrow C_2H_6\,(gas)$$

or by the disproportionation of two associatively adsorbed ethylene molecules:

$$2C_2H_4\,(a) \longrightarrow C_2H_2\,(a) + C_2H_6\,(gas)$$

The question is therefore raised as to the relative extents of associative and dissociative adsorption of ethylene. Magnetic studies using nickel–silica catalysts[9] suggest that at 0 °C ethylene is adsorbed associatively (1), and that

$$\underset{*\qquad\qquad *}{H_2C-CH_2}\quad \text{or}\quad \underset{*}{H_2C\!=\!CH_2}$$

(1)

as the temperature is increased dissociation of ethylene, resulting in the formation of a greater number of adsorbate–adsorbent bonds, and self-hydrogenation become important, until at temperatures above 130 °C carbon–carbon bond fission and the formation of surface carbide occur. These conclusions are supported by volumetric studies of ethylene adsorption on nickel[10] and on palladium.[11] At 0 °C the hydrogen : carbon ratio of the adsorbed species is about 1.5, falling to 1.0 at room temperature.

Field emission microscopy (FEM) studies of ethylene on iridium[12] and tungsten[13] also indicate that, as the temperature is raised, associative adsorption gives way to dissociation and the ultimate formation of surface carbide.

The co-existence of at least two modes of ethylene adsorption has been clearly demonstrated in studies of [^{14}C]ethylene adsorption at ambient temperature on a series of alumina-supported metals.[14] When [^{14}C]ethylene is adsorbed on to alumina-supported nickel, ruthenium, rhodium, palladium, iridium, or platinum, it can be shown that only a fraction of the initially adsorbed ethylene can be removed by hydrogenation, by molecular exchange with non-radioactive ethylene, or by evacuation. Furthermore, it is the same fraction which is removed by each of these processes. The remainder of the initially adsorbed ethylene is unaffected by these processes and is retained on

[5] G. I. Jenkins and E. Rideal, *J. Chem. Soc.*, 1955, 2490.
[6] O. Beeck, A. E. Smith, and A. Wheeler, *Proc. Roy. Soc.*, 1940, **A177**, 62.
[7] H. L. Pickering and H. C. Eckstrom, *J Phys. Chem.*, 1959, **63**, 512.
[8] O. Beeck, *Discuss Faraday Soc.*, 1950, **8**, 118.
[9] P. W. Selwood, *J. Amer. Chem. Soc.*, 1961, **83**, 2853.
[10] D. W. McKee, *J. Amer. Chem. Soc.*, 1962, **84**, 1109.
[11] S. J. Stephens, *J. Phys. Chem.*, 1958, **62**, 714.
[12] J. R. Arthur and R. S Hansen, *J Chem. Phys.*, 1962, **36**, 2062.
[13] R. S. Hansen and N. C. Gardner, *J. Phys. Chem.*, 1970, **74**, 3646.
[14] D. Cormack, S. J. Thomson, and G. Webb, *J. Catalysis*, 1966, **5**, 224.

the catalyst surface. The adsorptive capacity of the catalysts decreases in the order Ni > Rh > Ru > Ir > Pt > Pd, while the average percentage of the initially adsorbed ethylene retained by the surface is:

$$Pd > Ru > Ni \geqslant Rh > Ir > Pt$$
$$63.5 \quad 42.0 \quad 24.0 \quad 22.5 \quad 16.0 \quad 6.5$$

The retained species has been identified with a dissociatively adsorbed state of ethylene, although it is not possible to deduce the precise nature of the species from these studies. It is also difficult to correlate the extent of retention on a particular metal with other known catalytic properties, e.g. hydrocarbon cracking ability or hydrocarbon–deuterium exchange activity, of the metal in question.

The radioactive tracer approach has also been used to investigate the reactivity of the retained species at various temperatures with alumina-supported palladium, platinum, and rhodium[15,16] and alumina- and silica-supported platinum.[17] With palladium–alumina catalysts for a series of hydrocarbons the extent of retention increases in the order cyclopropane < alkenes < acetylene at all temperatures between 20 and 200 °C. With all the catalysts used, except platinum–silica, below 200 °C no exchange between adsorbed [^{14}C]ethylene and gas-phase ethylene is observed. With platinum–silica at room temperature some 25% of the initially retained [^{14}C]ethylene undergoes molecular exchange, although molecular exchange is not observed at higher temperatures. Treatment of the retained species with hydrogen results in the formation of small yields of methane at 200 °C with palladium and at 100 °C with platinum and rhodium. The interaction of acetylene with [^{14}C]ethylene-precovered surfaces does not result in any displacement of surface radioactivity, although even on [^{14}C]ethylene-saturated surfaces acetylene self-hydrogenation to ethane and ethylene is observed.

The results of these radioactive tracer studies are consistent with the retained species arising from the formation of multiply bonded hydrogen-deficient surface residues such as species (2)—(5).

Associatively bonded acetylene (2) has been proposed on the basis of volumetric studies,[18–20] although, in view of the lack of reactivity of the

---

[15] G. Taylor, S. J. Thomson, and G. Webb, *J. Catalysis*, 1968, **12**, 150.
[16] G. Taylor, S. J. Thomson, and G. Webb, *J. Catalysis*, 1968, **12**, 191.
[17] J. A. Altham and G. Webb, *J. Catalysis*, 1970, **18**, 133.
[18] See ref. 5, p. 185.
[19] See ref. 10, p. 185.
[20] G. A. Harvey, PhD. Thesis, University of Glasgow, 1973.

[^{14}C]ethylene-retained species towards acetylene, it seems unlikely to play a major role in retention on supported platinum, palladium, or rhodium. LEED studies of ethylene adsorption on the (111) face of platinum[21] suggest that the adsorbed ethylene occupies four sites, thereby supporting the postulate of species such as (3).

The adsorption of acetylene on palladium–alumina[15] and on silica- and alumina-supported platinum[22] exhibits the same general characteristics as those observed with ethylene. There are, however, some important differences. The extent of acetylene retention is substantially greater than that of ethylene retention. Furthermore, the amounts of acetylene retained by 'clean' and ethylene-precovered surfaces are identical. When compared with the high selectivity observed in acetylene hydrogenation over these catalysts,[23,24] indicating that acetylene can effectively exclude the readsorption of the reactive form of ethylene, this latter observation may be taken as evidence that the adsorbed states of acetylene and ethylene involved in retention and hydrogenation are different. It may also be concluded that the heterogeneity of a surface depends not only upon the surface itself but also upon the adsorbate under examination. A similar conclusion may be drawn from FEM studies of ethylene and acetylene adsorption on iridium and tungsten,[25,26] where it has been observed that ethylene adsorption occurs predominantly on the (111) planes whereas acetylene adsorbs most readily upon the (110) crystal planes. The greater degree of acetylene retention may also be a reflection of the known ability of this molecule to form surface polymers.[27]

[^{14}C]Propylene adsorption on platinum–alumina or platinum–silica[17] differs from both acetylene and ethylene adsorption in so far as, in the temperature range 20—200 °C, a fraction of the initially retained [^{14}C]propylene will undergo molecular exchange with gas-phase propylene. This fraction is 35% with platinum–alumina and 26% with platinum–silica. The greater ease of exchange of propylene relative to ethylene suggests that a different type of adsorbed species may be in part responsible for propylene retention. With propylene surfaces species of the π-allyl type (6) may be formed, which are not possible with ethylene.

$$H_2C-CH-CH_2$$
$$\overline{|}$$
$$*$$

(6)

Further evidence for this proposal comes from the observation that ethylene

---

[21] D. L. Smith and R. P. Merril, *J. Chem. Phys.*, 1970, **52**, 5861.
[22] J. A. Altham and G. Webb, unpublished work.
[23] G. C. Bond and P. B. Wells, *J. Catalysis*, 1965, **4**, 211.
[24] G. C. Bond and P. B. Wells, *J. Catalysis*, 1966, **5**, 65, 419.
[25] See ref. 12, p. 3.
[26] See ref. 13, p. 3.
[27] L. H. Little, N. Sheppard, and D. J. C. Yates, *Proc. Roy. Soc.*, 1960, **A249**, 242.

adsorption can still occur on surfaces which have been effectively saturated with propylene, although the extent of ethylene adsorption is less than on a 'clean' surface.

The use of radioactive tracers has clearly demonstrated that the surfaces of metal catalysts are, as expected from measurements of heats of adsorption,[28] heterogeneous for hydrocarbon adsorption, although the degree of heterogeneity depends upon both the surface and, to some extent, the adsorbate molecule. This latter point is demonstrated from comparisons of the behaviour of [^{14}C]ethylene precovered palladium–alumina and [^{14}C]ethylene-precovered palladium films towards acetylene. In the former case no surface radioactivity is displaced by acetylene at temperatures between 20 and 200 °C. However, when a molecular beam of acetylene is allowed to interact with a [^{14}C]ethylene-precovered palladium film in the temperature range 15—90 °C ^{14}C-containing species were displaced from the surface in a two-stage process.[30] The first stage was believed to correspond to the displacement of C_2-species, whereas in the second slower stage C_4-hydrocarbons were displaced from the catalyst surface. These studies also demonstrate that only a fraction of the initially adsorbed hydrocarbon is active in catalysis, this fraction varying from metal to metal. Such observations suggest that comparisons of the activities of different metals for hydrocarbon hydrogenation,[2, 29] where the activity is based upon the metal surface area as determined by, for example, carbon monoxide adsorption, or from particle size distributions, may only be approximate and should be treated with some caution.

3 Infrared Spectra of Adsorbed Hydrocarbons

Although the radiotracer approach permits the direct observation of the fates of adsorbed species under reaction conditions,[30] it yields little or no *direct* information regarding the chemical identity of the various adsorbed species. Such information may, in principle, be obtained from i.r. spectroscopic studies of adsorbed hydrocarbons. Most of the i.r. studies have been carried out using silica-supported metal catalysts. The choice of silica lies in the large i.r. window which this material exhibits.[31] Extensive studies of the adsorption of ethylene and acetylene on silica-supported nickel, palladium, and platinum have been made.

The interpretation of the spectra of adsorbed hydrocarbons relies to a great extent upon comparisons with the spectra of model compounds. This has in some cases led to difficulties because of the lack of suitable transition-metal–hydrocarbon compounds. The recent studies of Sheppard and Ward,[32] who

---

[28] G. C. Bond, 'Catalysis by Metals', Academic Press, London and New York, 1962.
[29] G. C. A. Schuit and L. L. van Reijen, *Adv. Catalysis*, 1958, **10**, 242.
[30] S. J. Thomson and J. L. Wishlade, *Trans. Faraday Soc.*, 1962, **58**, 1170; see also ref. 14, p. 185; J. J. McCarroll and S. J. Thomson, *J. Catalysis*, 1970, **19**, 144.
[31] L. H. Little, 'Infra-red Spectra of Adsorbed Species', Academic Press, London and New York, 1966.
[32] N. Sheppard and J. W. Ward, *J. Catalysis*, 1969, **15**, 50,

used a series of substituted hydrocarbons and organometallic compounds as models, and of Morrow,[33] who used the spectra of platinum alkyl complexes for comparison, have largely overcome this problem. Sheppard[34] has indicated that one of the main limitations of the i.r. approach to studying adsorbed hydrocarbons is that the presence of surface intermediates may not be revealed if the appropriate band intensities are weak. This problem has also been discussed by Avery,[35] who postulated that C—H bands associated with adsorbed carbon atoms (those directly involved in carbon–metal bond formation), are too weak to be observed, and that the observed spectra arise from either terminal C—H groups or C—H groups neighbouring the adsorbed carbon atoms. In spite of these difficulties, this approach has contributed significantly to our understanding of the nature and reactivity of the adsorbed states of unsaturated hydrocarbons.

The adsorption of ethylene on a 'bare' platinum–silica catalyst[36-38] at room temperature leads to spectra in which the associatively bonded structure (1) is predominant. Bonds ascribable to the dissociatively adsorbed species (2) are also observed. On admission of hydrogen to the ethylene-precovered surface ethane is observed in the gas-phase, and the spectrum increases in intensity. This has been interpreted as indicating the formation of surface carbides in the initial adsorption. On increasing the temperature to 95 °C, bands corresponding to surface n-butyl groups are observed together with the appearance of butane in the gas-phase. It was suggested that these C_4-species were the result of random polymerization of dissociatively adsorbed ethylenic residues. However, no redistribution of carbon-13 is observed in the reaction of preadsorbed [1-^{13}C]ethylene with hydrogen[39] and, together with thermal desorption data[40] for the ethylene–platinum system, this may be taken to indicate that on platinum the retained species is a discrete C_2-unit.

The spectra of ethylene adsorbed on to nickel–silica show close similarities to those observed with platinum–silica. The major difference between the two catalysts is the temperature range over which the various bands are observed. Thus, with nickel–silica, surface n-butyl groups appear at room temperature whereas these only become apparent with platinum–silica above 95 °C. In general, there appears to be a greater degree of dissociation, yet a lower retention, with platinum than with nickel. But-1-ene shows similar behaviour.[38] With nickel–silica it is likely that the retained species are C_4-species rather than C_2-species observed with platinum.[38,41]

Adsorption of ethylene on 'bare' palladium–silica produces only weak

[33] B. A. Morrow, *Canad. J. Chem.*, 1970, **48**, 2192.
[34] N. Sheppard, *Discuss Faraday Soc.*, 1966, **41**, 254.
[35] N. R. Avery, *J. Catalysis*, 1970, **19**, 15.
[36] B. A. Morrow and N. Sheppard, *J. Phys. Chem.*, 1966, **70**, 2406.
[37] N. Sheppard, *Discuss. Faraday Soc.*, 1966, **41**, 177.
[38] B. A. Morrow and N. Sheppard, *Proc. Roy. Soc.*, 1969, **A311**, 391.
[39] B. A. Morrow, *J. Catalysis*, 1969, **14**, 279.
[40] R. Komers, Y. Amenoyima, and R. J. Cventanovic, *J. Catalysis*, 1969, **15**, 293.
[41] J. B. Peri, *Discuss. Faraday Soc.*, 1966, **41**, 121.

i.r. bands.[42] These bands indicate the presence of species with double-bond character [structure (2)] together with surface methyl and methylene groups. On admission of hydrogen the C=C bands disappear and intense bands corresponding to surface ethyl groups are observed.

Acetylene adsorption follows a similar pattern to that observed with ethylene. Thus, on silica-supported nickel, palladium, or platinum,[32] adsorption of acetylene on the 'bare' metal gives rise to bands ascribable to olefinic species [structure (2)] and to surface alkyl groups. Addition of hydrogen results in an intensification of the spectra and the appearance of new bands corresponding to surface alkyl groups of average structure $CH_3(CH_2)_n$, where $n \geqslant 4$ for platinum and $n = 3$ for nickel. As with ethylene, the degree of dissociation of the adsorbed acetylene is greater with platinum than with nickel, whereas palladium is very similar to platinum except that on the former catalyst the proportion of olefinic C—H bonds is somewhat higher. Recent studies of acetylene adsorption on palladium–alumina[43] and γ-alumina-supported platinum,[44] and on explosively dispersed copper, palladium, and nickel[45] show that the spectra are very similar to those obtained with the silica-supported metals.

Avery[35] has described spectra for a range of branched-chain and linear olefins adsorbed on to platinum–silica catalysts. These show that, with branched-chain olefins, the predominant species are saturated adsorbed alkyl species, whereas with linear olefins, which give weaker initial absorption bands, the tendency to form dehydrogenated residues is much greater.

In general the study of i.r. spectra of adsorbed olefins and acetylenes leads to similar conclusions to those drawn from radiochemical studies. The main difference between the two lies in the ability of the former technique to detect surface polymers, particularly dimers of ethylene and acetylene. Further evidence for the dimerization of ethylene at step edges and (111) terraces of nickel has been obtained by FEM.[46] Kokes[47] has also observed the formation of dimers during ethylene hydrogenation over cobalt, and thermal desorption studies indicate that both ethylene and acetylene undergo polymerization on alumina- and silica-supported platinum catalysts[48] and on palladium films. The polymerization of acetylene on metal catalysts is also a well-established phenomenon.[49]

From the foregoing discussion it may be concluded that, when unsaturated hydrocarbons interact with a metal surface, both dissociatively and associatively adsorbed species are formed. For a particular hydrocarbon the relative

[42] L. H. Little, 'Infra-red Spectra of Adsorbed Species', Academic Press, London and New York, 1966.
[43] N. P. Sokolova, L. A. Kazakova, and L. A. Borisenko, *Russ. J. Phys. Chem.*, 1970, **44**, 1515.
[44] S. S. Randhava and A. Rehmat, *Trans. Faraday Soc.*, 1970, **66**, 235.
[45] C. P. Nash and R. P. DeSieno, *J. Phys. Chem.*, 1965, **69**, 2139.
[46] L. Whalley, B. J. Davis, and R. L. Moss, *Trans. Faraday Soc.*, 1970, **66**, 3143.
[47] R. J. Kokes, *J. Catalysis*, 1969, **14**, 83.
[48] J. A. Altham, PhD. Thesis, University of Glasgow, 1969.
[49] J. Sheridan, *J. Chem. Soc.*, 1944, 373; see also refs. 23 and 24.

proportions of each species vary from metal to metal, and for a particular metal the proportions depend on the structure of the metal surface. This last point is well illustrated in a study of the adsorption and hydrogenation of [^{14}C]acetylene on a cold-worked palladium foil,[50] in which it was demonstrated that the catalytic activity of the latter shows a marked variation with annealing. [^{14}C]Acetylene adsorption studies demonstrate the existence of four types of adsorbed acetylene: A, which undergoes desorption on evacuation; B, which is removed from the surface during hydrogenation; C, which is not removed during hydrogenation, but is removed by treatment with hydrogen at 150 °C; and D, which remains on the surface after reduction at 150 °C. The variation in the relative amounts of these four types of adsorbed state with temperature of annealing of the worked foil is shown in Figure 1.

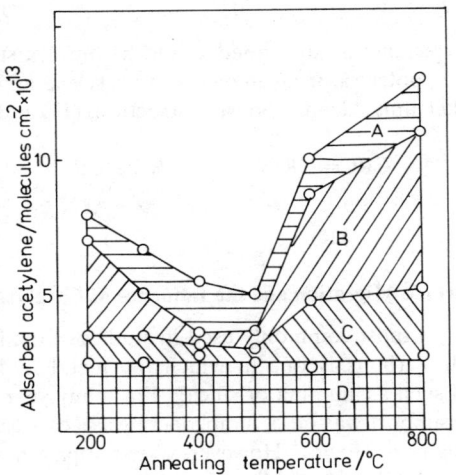

Figure 1 *Variation in the fractions of various forms of adsorbed acetylene on palladium foil with temperature of annealing.*
(Reproduced by permission from *J. Phys. Chem.*, 1969, **73**, 1618)

It was also considered that the type B species are the species responsible for hydrogenation and that these species can occupy two kinds of site (I) and (II). Site (I) can be identified with lattice imperfections in the (110) crystal plane which disappear on annealing at temperatures between 200 and 300 °C. Site (II) can be correlated with the lattice planes or boundaries which are preferentially developed during the disappearance of the (110) planes and the growth of (111) planes at annealing temperatures of around 600 °C.

The effect of addition of hydrogen to a hydrocarbon-precovered surface may be taken as a clear indication that the species active in hydrogenation is

---

[50] Y. Inoue and I. Yasumori, *J. Phys. Chem.*, 1969, **73**, 1618.

an associatively bonded species, as has been assumed from mechanistic studies of hydrogenation reactions.[49] There still exists, however, some doubt as to the type of surface–adsorbate bonding involved in the associatively adsorbed species. From i.r. studies it has been concluded that with olefins a di-σ-bonded surface complex (7) is formed, whereas, following the suggestions of Rooney et al.,[51,52] a so-called π-complex (8) has been assumed in hydrogenation studies. These differences are perhaps not too significant since, by analogy with the bonding in transition-metal–olefin and metal–acetylene complexes, the bonding in structure (8) will have both a σ- and a π-component, and (8) may be better represented as a π-complex[53] with structure (9). With

```
     RHC—CHR           RHC=CHR          RHC—CHR
     /     \               |             \   /
    M       M              M              M
       (7)               (8)              (9)
```

acetylene the corresponding di-σ-bonded and π-complex structures are (10) and (11). Surface potential measurements of acetylene and ethylene chemisorbed on to nickel films[54] lend support to structures (11) and (8), respectively.

```
     RC=CR              RC≡CR
     /   \               /\
    *     *             M  M
     (10)               (11)
```

4 Surface Migration and the Influence of Catalyst Supports

The type of catalyst most commonly used in studies of the hydrogenation of unsaturated hydrocarbons is one in which the metal is dispersed upon a support material such as alumina or silica. For many years it was generally thought that the support was inert, simply acting as a carrier for the metal in a finely divided, high area form. However, over the past few years there has been a growing body of evidence that the support may not be inert but may influence the catalytic activity in any of several ways. First, it may interact chemically with the metal thereby modifying the properties of the latter. Such reasons have been invoked to explain the modifications in electrical conductivity when a metal is dispersed on different supports,[55] the differences in the activities and selectivities for benzene hydrogenation of platinum when supported on polyamides, molecular sieves, and alumina,[56] and the differences in activity for buta-1,3-diene hydrogenation of γ-alumina, boehmite, and silica-supported gold catalysts. Recently, Figueras et al.[57] have shown that

---

[51] J. Rooney, *J. Catalysis*, 1963, **2**, 53.
[52] J. Rooney and G. Webb, *J. Catalysis*, 1964, **3**, 488.
[53] P. S. Braterman and R. J. Cross, *Chem. Soc. Rev.*, 1973, **2**, 271.
[54] J. C. P. Mignolet, *Discuss. Faraday Soc.*, 1950, **8**, 105.
[55] S. J. Thomson and G. A. Harvey, *J. Catalysis*, 1971, **22**, 359.
[56] P. Dini, D. Dones, S. Montelatchi, and N. Giordano, *J. Catalysis*, 1973, **30**, 1.
[57] F. Figueras, R. Gomez, and M. Primet, *Adv. Chem. Ser.*, 1973, **121**, 480.

the catalytic activity of supported palladium for benzene hydrogenation may vary depending on the Lewis acid properties of the support. Increasing Lewis acidity of the support results in an increase in the activity per surface metal atom. The authors interpret their observations in terms of the modification of the electronic state of palladium due to metal-oxidizing site interactions. Second, the support may exert an influence on the actual structure of the metal,[58] although such effects may be difficult to separate from the chemical effects. Third, the support may act either as a source of reactive intermediates or as a seat of reaction, through the migration of adsorbed species between metal and support.

The phenomenon of surface migration between metal and support, which has been termed 'spill-over',[59] and which has recently been reviewed,[60] has been shown to be of particular importance with hydrogen on supported metals.[61-63] Most of these studies have been concerned with the spill-over of hydrogen from metal to support. Evidence for the back-migration of hydrogen from support to metal is less well documented. The most direct evidence for hydrogen back-migration has been obtained by Bond et al.[64] who used an alkene titration method to characterize a number of supported metal catalysts. Alkene titration of a platinum–H_xWO_3 catalyst at 100 °C yields hydrogen equivalent to a hydrogen : platinum ratio ($n_H : n_{Pt}$) of approximately 277. Since the support alone does not hydrogenate the alkene at this temperature back-migration of hydrogen must be responsible for the hydrogenation. Similar observations have been made with platinum–silica and platinum–alumina catalysts[17] using a radiotracer approach to monitor the extent of hydrogen spill-over. In these studies it was shown by tritium exchange that the hydrogen associated with the hydroxy-groups on the surface of alumina or silica can undergo reaction with ethylene or acetylene in the temperature range 20—350 °C. Under the conditions used, the direct exchange between the hydrocarbon and the support hydroxy-groups, in the absence of platinum, is not observed, thus indicating the back-migration of hydrogen from support to metal. It is worth noting, however, that some care must be exercised in the interpretation of the results of such studies in terms of hydrogen spill-over, since it has also been established that hydrogen, in a reactive form, may be occluded in the metal as well as adsorbed on the surface of the catalyst. This occlusion of hydrogen has been the subject of studies by Wells et al.[65] of butene isomerization over Group VIII metal powders. It

---

[58] D. A. Buchanan and G. Webb, unpublished results.
[59] M. Boudart, M. A. Vannice, and J. E. Benson, *Z. Phys. Chem.*, 1969, **64**, 171.
[60] P. A. Sermon and G. C. Bond, *Catalysis Rev.*, 1973, **8(2)**, 211.
[61] J. E. Benson, H. W. Kohn, and M. Boudart, *J. Catalysis*, 1966, **5**, 307.
[62] J. H. Sinfelt and P. J. Lucchesi, *J. Amer. Chem. Soc.*, 1963, **85**, 3365.
[63] K. H. Sancier, *J. Catalysis*, 1971, **20**, 106.
[64] G. C. Bond, P. A. Sermon, and J. B. P. Tripathi, Paper presented at Louvain meeting of Societe Chemique de Belgique, Sept. 1972.
[65] S. D. Mellor, N. C. Smith, and P. B. Wells, ref. 48 in P. B. Wells, 'Surface and Defect Properties of Solids', ed. M. W. Roberts and J. M. Thomas, (Specialist Periodical Reports), The Chemical Society, London, 1971, Vol. 1, p. 236.

has also been observed with platinum blacks[66] and with palladium black.[67] Clearly, before hydrogen spill-over and back-migration can be conclusively proved, it is necessary to show that the amounts of hydrogen involved are greater than can be ascribed to direct adsorption on the metal and occlusion within the metal.

In contrast with the apparent wealth of information regarding hydrogen spill-over, the migration of adsorbed hydrocarbons between metal and support, in the context of hydrogenation reactions, has received little attention. The migration of hydrocarbons has, however, often been invoked to explain the mechanism of bifunctional catalysis.[68] Evidence for the participation of the support in the reactions of olefins and acetylenes over supported metals has been obtained by the use of catalyst poisons. Webb and Macnab have studied the hydrogenation, hydroisomerization and deuterium exchange of but-1-ene over rhodium–silica[69] in the temperature range 0—69 °C. The catalysts were progressively poisoned with mercury, the extent of mercury coverage being measured by a radioactive counting technique and the rates of the various reactions determined. The variations in rates with mercury coverage are shown in Figure 2. It was suggested that the results are satisfactorily interpreted by considering that hydrogenation and but-1-ene exchange are confined to the metal, whereas isomerization occurs on the support and involves the migration of hydrocarbon from the metal to the support. A similar mechanism might be invoked to explain the observation that alumina, when used as a metal catalyst support, assists the isomerization of n-butenes over Group VIII metals in the absence of hydrogen.[70]

Reid *et al.* have studied the adsorption of ethylene[71] and of acetylene[72] on 'bare' and carbon monoxide precovered rhodium–silica and rhodium–alumina catalysts using ^{14}C-tracer techniques. When adsorbed on to a 'clean' surface both ethylene and acetylene are adsorbed in a two-stage process; a fast primary region, followed by a slower secondary region. From carbon monoxide adsorption it was shown that the change from primary to secondary adsorption corresponds to carbon monoxide monolayer coverage (see Figure 3). Further, on carbon monoxide precovered surfaces, adsorption of ethylene or acetylene in the secondary region could still occur. On the assumption that carbon monoxide adsorption is restricted to the metal it was concluded that the primary adsorption of ethylene and acetylene corresponds to adsorption on the metal; the principal secondary species are formed by migration of hydrocarbon from the metal to support. Both direct adsorption

---

[66] Z. Paal and S. J. Thomson, *J. Catalysis*, 1973, **30**, 96.
[67] L. V. Babenkova, N. M. Popova, D. V. Sokol'skii, and V. K. Solnyserkova, *Doklady Akad. Nauk S.S.S.R.*, 1973, **210**, 888.
[68] G. A. Mills, H. Heinemann, T. H. Milliken, and A. G. Oblad, *Ind. and Eng. Chem.*, 1953, **45**, 135.
[69] G. Webb and J. I. Macnab, *J. Catalysis*, 1972, **26**, 226.
[70] P. B. Wells and G. R. Wilson, *J. Catalysis*, 1967, **9**, 70.
[71] J. U. Reid, S. J. Thomson, and G. Webb, *J. Catalysis*, 1973, **29**, 421.
[72] J. U. Reid, S. J. Thomson, and G. Webb, *J. Catalysis*, 1973, **30**, 372.

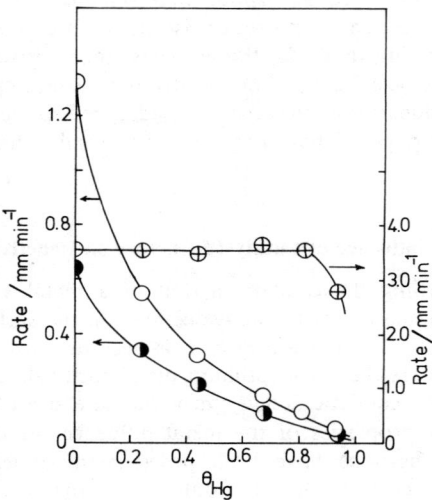

Figure 2 *The effect of mercury coverage (θ_{Hg}) upon the rates of hydrogenation (○), isomerization (⊕) and olefin exchange (◐) for the reaction of but-1-ene with deuterium over rhodium–silica at 48 °C*
(Reproduced by permission from *J. Catalysis*, 1972, **26**, 226)

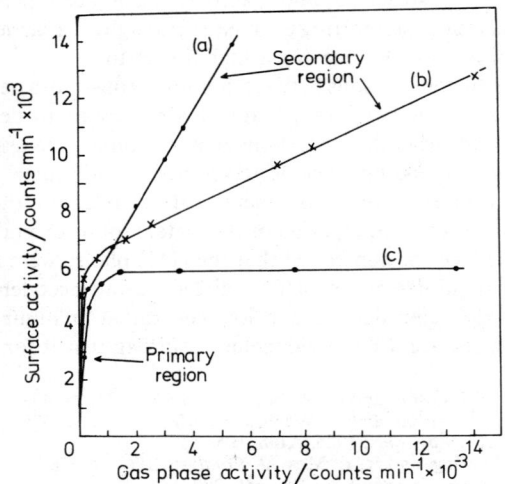

Figure 3 *Adsorption isotherms for* (a) $^{14}C_2H_4$, (b) $^{14}C_2H_2$, *and* (c) ^{14}CO *on rhodium–alumina at* 20 °C
(Reproduced by permission from *J. Catalysis*, 1973, **29**, 421; 1973, **30**, 372)

on the support and physical adsorption were ruled out on the basis that the gradients of the secondary adsorption isotherms are substantially greater than the corresponding values for the supports alone. It was also concluded that, while both associatively and dissociatively adsorbed species are responsible for primary adsorption, the secondary adsorbed species are associatively bonded to the surface. These conclusions are substantiated by thermal desorption studies.[73,74]

5 The Influence of Catalyst Structure on Reactivity

Any theory regarding the catalytic action of a metal surface requires a specification of the nature of the active centres on the surface. In order to gain a knowledge of the active centres it is necessary to obtain answers to such questions as: Is there an optimum metal–metal distance for catalytic activity? Does the catalytic activity vary with the size of metal particles? How do the bulk properties of the metal influence the catalytic activity? Although these questions have been posed many times since Taylor[75] postulated his active-site theory, it is only during the past decade, with the development of techniques for the accurate physical characterization of small metal particles,[76] that systematic examinations of the influence of catalyst structure upon catalytic activity have been undertaken. Theoretically it is expected that the most efficient use of metals as catalysts should be achieved when the metal is in a finely divided state, thereby making most of the metal atoms accessible to reactants. It is not surprising therefore that most studies of the influence of structure upon activity have been performed using supported metals, since supporting the metal on a high area carrier is the easiest and most practical way of achieving a high dispersion.

Bond[77] has used comparisons between homogeneously and heterogeneously catalysed interconversions of unsaturated hydrocarbons to deduce that the reactive state of adsorbed hydrocarbons may reasonably be assumed to be a π-complex. On this assumption he developed a molecular orbital model appropriate to a face-centred cubic metal. By considering the direction of emergence and degree of occupation of the metal atomic orbitals at the (100), (110), and (111) faces he concluded that the (111) planes were least suited to the adsorption requirements of olefins, diolefins, and acetylenes. Poltorak and co-workers[78-80] considered the various co-ordination numbers of the surface atoms of a face-centred cubic metal, assuming that it forms octahedral

[73] J. U. Reid, S. J. Thomson, and G. Webb, *J. Catalysis*, 1973, **29**, 433.
[74] J. U. Reid, S. J. Thomson, and G. Webb, *J. Catalysis*, 1973, **30**, 378.
[75] H. S. Taylor, *Proc. Roy. Soc.*, 1925, **A108**, 105.
[76] See *e.g.* T. A. Dorling, and R. L. Moss, *J. Catalysis*, 1967, **7**, 378.
[77] G. C. Bond, *Discuss. Faraday Soc.*, 1966, **41**, 200.
[78] O. M. Poltorak and V. S. Boronin, *Russ. J. Phys. Chem.*, 1965, **39**, 1329.
[79] O. M. Poltorak and V. S. Boronin, *Russ. J. Phys. Chem.*, 1966, **40**, 1436.
[80] O. M. Poltorak, V. S. Boronin, and A. N. Mitrofanova, Proceedings 4th International Congress on Catalysis, Akademiai Kiado, Budapest, 1971, **2**, 276.

crystallites with regular faces. The distribution of surface atoms of different co-ordination number as a function of crystallite size was also deduced. It was concluded that for crystallites in the range 10—20 Å a large number of atoms of abnormally low co-ordination number are present, whereas in the 20—50 Å region, there is a preponderance of atoms with co-ordination numbers corresponding to crystal edges and faces. Thus, by determining the specific activities of catalysts with particle size distributions in the range 10—50 Å, it should be possible to establish relationships between activity and average co-ordination number of a surface metal atom. A similar conclusion has been reached by Bond[81] and by van Hardeveld and van Montfoort[82] from considerations of the nature of sites present at the surface of both complete and incomplete cubo-octahedral crystallites, particular attention being paid to the numbers of so-called B_5 sites.

From these theoretical considerations it might be expected that the catalytic activity of a metal would vary substantially with particle size. The most surprising feature of the quantitative measurements so far made on the activity of small metal particles is the apparent lack of sensitive variation of activity with particle size in most systems. Thus, the specific activities of a series of platinum–silica catalysts for hex-1-ene hydrogenation were observed to be independent of particle size,[83] whereas over an increase in particle size of two orders of magnitude no change in specific activity was observed in the hydrogenation of benzene over platinum–silica catalysts.[84] Similarly, Boudart et al.[85] have shown that the specific activities of highly dispersed platinum on alumina and bulk platinum are virtually identical for cyclopropane hydrogenation. These observations have led Boudart to classify reactions as facile and demanding.[86] It would appear that the majority of hydrogenation reactions of unsaturated hydrocarbons fall into the former class. However, from the studies of van Hardeveld and Hartog[87] on benzene hydrogenation over nickel and iridium catalysts, even this generalization may have to be limited to specifying particular metals. Clearly the situation is extremely complex and many more studies need to be made before any definitive relationship between catalytic activity and metal particle size can be established. Such studies will also need to take into account factors such as the perturbation of surface structure due to carbiding,[86] the influence of the method of preparation of the catalyst,[88] and the actual structure of the real metal crystallites in terms of lattice defects and imperfections.

[81] G. C. Bond, Proceedings 4th International Congress on Catalysis, Akademiai Kiado, Budapest, 1971, **2**, 266.
[82] R. van Hardeveld and A. van Montfoort, *Surface Sci.*, 1966, **4**, 396; 1969, **17**, 90
[83] O. M. Poltorak and V. S. Boronin, *Russ. J. Phys. Chem.*, 1965, **39**, 781.
[84] T. A. Dorling and R. L. Moss, *J. Catalysis*, 1966, **5**, 111.
[85] M. Boudart, A. Aldag, J. E. Benson, N. A. Dougharty, and C. G. Harkins, *J. Catalysis*, 1966, **6**, 92.
[86] M. Boudart, *Adv. Catalysis*, 1969, **20**, 153.
[87] R. van Hardeveld and F. Hartog, *Adv. Catalysis*, 1972, **22**, 75.
[88] R. G. Oliver and P. B. Wells, Proceedings 5th International Congress on Catalysis, North-Holland Publishing Co., Amsterdam, 1973, **1**, 659.

Erratum

Volume 2, 1973

Title page: It is regretted that the following name and address was omitted from the list of Reporters:

D. A. Whan, *University of Edinburgh*

Author Index

Acton, A. F., 102
Adams, L. H., 151
Aldag, A., 197
Allan, C. J., 169
Allan, P. 96, 107
Allison, D. A., 169
Allison, J., 169
Allpress, J. G., 3, 5, 12, 18, 44,
Altham, J. A., 186, 187, 190
Amelinckx, S., 14, 21
Amenoyima, Y., 189
Anderson, J. S., 2, 15, 21, 22, 25, 41, 49
Andersson, G., 33
Andersson, S., 36
Andreasen, A. S., 51
Andrews, E. H., 131
Anstis, G. R., 5
Anthony, K. H., 58, 59, 72
Åsbrink, S., 33
Argon, A. S., 126
Armstrong, R. W., 61, 89, 92
Arthur, J. R., 185, 187
Avery, N. R., 189

Babenkova, L. V., 194
Bambynek, W., 158
Banfield Younghusband, H., 76
Baun, W. L., 168
Bearden, J. A., 169
Beavis, L. C., 164
Beeck, O., 185
Beevers, R. B., 62
Ben-Abraham, S. J., 59
Benson, G. C., 146
Benson, J. E., 193, 197
Bent, H. A., 132
Bevan, D. J. M., 22, 55
Bevis, M., 96, 102, 106, 107, 116
Bilby, B. A., 72, 106
Blasse, G., 12
Blundell, D. J., 97
Böhm, J., 169
Bond, G. C., 184, 187, 188, 190, 193, 196, 197
Borisenko, L. A., 190
Boronin, V. S., 196, 197
Boudart, M., 193, 197
Bouligand, Y., 59, 61, 86, 88
Bowden, P. B., 126
Bradshaw, A. M., 155, 162, 169, 172, 175, 180
Bragg, W. L., 75

Braterman, P. S., 192
Breedon, J. E., 96
Brinkman, H., 156
Brouwer, W. S., 55
Browne, J. M., 22, 41
Bruining, H., 154
Buchanan, D. A., 184, 193
Bullough, R., 72
Burbank, R. D., 97
Burr, A. F., 156, 169
Bursill, L. A., 12, 21, 25, 34, 36

Cano, R., 88
Caroli, C., 88
Caspar, D. L. D., 63
Chamberlain, M. B., 168
Chang, E. P., 95
Chang, J. J., 175
Cheetham, A. K., 22, 39
Chau, T.-W., 61, 72
Christian, J. W., 99
Christoffersen, R. E., 152
Chu, S. Y., 152
Cladis, P. E., 87, 88
Codling, K., 177
Cohen, R. L., 179
Condon, E. U., 157
Conklin, J. B., 181
Cook, D. B., 152
Cormack, D., 185, 188
Cowley, J. M., 4, 10, 18
Craseman, B., 158
Crellin, E. B., 96
Crocker, A. G., 102, 106
Cross, R. J., 192
Cventanovic, R. J., 189

Dafermos, C. M., 87
Darling, T. A., 196, 197
Das, E. S. P., 72, 89, 92
Davis, B. J., 190
de Gennes, P. G., 87, 88
Delavignette, P., 14
DeSieno, R. P., 190
Dev, B., 156
de Wit, R., 58, 61, 65, 66, 89, 92
Dini, P., 192
Dones, D., 192
Dooley, G. J., 164
Dougharty, M. A., 197
Dubois-Violette, E., 88
Dundurs, J., 72, 92
Dyaloshinskii, I. E., 88

Eckstrom, H. C., 185
Eiblei-Machtler, R., 181
Ericksen, J. L., 88

Erickson, W. D., 141, 143, 150
Ern, V., 181
Ertl, G., 178
Eshelby, J. D., 59, 61
Essmann, U., 58
Eujen, H. M., 142

Fabian, D. J., 156
Fahlman, A., 180
Fan, C., 88
Faulds, H., 62
Faulkner, J. S., 178
Fellenzer, H., 177
Figueras, F., 192
Fink, R. W., 158
Foster, A. J., 165
Ford, B., 152
Frank, F. C., 58, 61, 93, 126
Freund, H. U., 158
Friberg, S., 33
Friedel, G., 62
Friedel, J., 58, 61, 87
Frost, A. A., 132, 136, 152
Fryxell, G. A., 77
Fuller, R. B., 61, 63

Gadó, P., 12
Galligan, J. M., 59, 61
Galton, F., 62
Galy, J. L., 55
Gardner, N. C., 185, 187
Geil, P. H., 95, 97
Gelius, U., 169
Gibb, R. M., 15, 25
Gilman, J. J., 59
Giordano, N., 192
Gleiter, H., 126
Gomez, R., 192
Grandjean, F., 62, 88
Grant, J., 184
Gray, R. W., 95
Green, L. G., 143
Grey, I. E., 32, 33
Grinton, G. R., 4
Groves, G. W., 112
Grubb, D. T., 112
Gruehn, R., 41, 42
Gruler, H., 87
Günther, H., 89
Gunn, S. R., 143
Gupta, V. B., 126

Haas, T. W., 164
Hall, G. G., 152
Hamrin, K., 180
Hansen, R. S., 185, 187
Harris, W. F., 59, 63, 65, 75, 76, 77, 78, 89

199

Hartog, F., 197
Harvey, G. A., 186, 192
Hasiguti, R. R., 89
Hasle, G. R., 77
Hara, T., 94, 95
Harkins, C. G., 197
Hay, I. L., 95
Head, A. K., 6, 77
Heinemann, H., 194
Hewat, E. A., 5
Hills, G. J., 76
Hirth, J. P., 58, 59
Holland, V. F., 119, 126
Holtzberg, F., 41
Houston, J. E., 153, 155, 158, 159, 162, 166, 168, 172, 175, 179
Huang, W., 72
Hüfner, S., 179
Hull, R., 76
Hutchinson, J. L., 18, 22, 41
Hyde, B. G., 12, 21, 25, 34, 36

Iijima, S., 18, 20, 22, 48
Imura, H., 88
Inoue, Y., 191
Ishida, Y., 75
Ishii, K., 116

Jackson, J. F., 96
Johanssen, G., 169, 180
Jenkins, G. I., 185, 186

Källne, E., 180
Kanski, J., 155, 160, 161, 176
Kato, S., 155
Kaye, G. W. C., 151
Kazakova, L. A., 190
Kellenberger, E., 76
Keller, A., 93, 95, 96, 97, 112
Kiho, H., 95, 116
Kimball, G. E., 132
Kimura, S., 42
Kirschner, J., 155, 175, 180
Kiselev, N. A., 76
Kléman, M., 58, 59, 61, 86, 87, 88
Klug, A., 63, 76
Kohn, H. W., 193
Kokes, R. J., 190
Komers, R., 189
Kovacs, A. J., 97
Kröner, E., 58, 59
Kroupa, F., 72
Kuo, H. H., 72, 92

Laby, T. H., 151
Langreth, D. C., 175
Laramore, G. E., 155, 175
Laver, W. G., 76
Leder, L. B., 162
Lehmann, O., 62
Lejček, L., 72
Levin, E. M., 44
Lewis, D., 126
Li, J. C. M., 59, 72

Liberman, D. S., 101
Liefeld, R. J., 155, 156, 157
Lincoln, F. J., 18, 22
Linnett, J. W., 141, 143, 146, 152
Little, L. H., 187, 188, 190
Liu, G. C. T., 72
Lohneiss, H., 175
Lomer, W. M., 75
Long, R. L., 164
Loopstra, B. O., 51
Lothe, J., 58
Love, A. E. H., 65
Lu, T.-L., 72
Lucchesi, P. J., 193
Lynch, D. F., 5

McCaffrey, J. W., 175
McCarroll, J. J., 188
McCrum, N. G., 95
McKee, D. W., 185
Macnab, J. I., 194
Maddams, W. F., 126
Maggiora, G. M., 152
Magnéli, A., 12, 33
Mann, A. W., 55
Marcinkowski, M. J., 72, 89, 92, 96
Mark, H., 158
Markham, R., 76
Marks, R. W., 61
Marshall, C. A. W., 156
Mauguin, C., 62
Meier, G., 87
Mellor, S. D., 193
Menzel, D., 162, 169, 172
Merril, R. P., 187
Meyer, R. B., 87
Mignolet, J. C. P., 192
Miller, R. L., 126
Milliken, T. H., 194
Mills, G. A., 194
Mitchell, L. H., 77
Mitrofanova, A. N., 196
Montelatchi, S., 192
Moodie, A. F., 4, 5, 10
Morimoto, N., 56
Morrow, B. A., 189
Morrow, D. R., 95
Moser, R., 53
Moss, R. L., 190, 196, 197
Moyes, R. B., 184
Mumme, W. G., 33
Mura, T., 72, 92
Murthy, M. S., 164
Musket, R. G., 164, 168

Nabarro, F. R. N., 58, 61, 76, 92
Nagel, D. J., 175
Nakazawa, H., 56
Nash, C. P., 190
Neckel, A., 181
Nehring, J., 88
Nelson, J. L., 152
Neurath, A. R., 76
Newman, B. A., 95
Nilsson, P. O., 155, 160, 161, 176

Nimmo, K. M., 22
Nishiguchi, K., 56
Nishino, M., 61
Nordling, C., 180
Nye, J. F., 75

Oblad, A. G., 194
O'Connor, A., 93
Okano, K., 88
O'Keefe, M. A., 5, 36
Oliver, R. G., 197
Orsay Liquid Crystal Group, 88
Oseen, C. W., 62

Paal, Z., 194
Packer, J. C., 152
Pakiari, A. H., 137
Palmberg, P. W., 162
Papaconstantopoulos, D. A., 175
Park, R. L., 153, 155, 158, 159, 162, 166, 168, 172, 175, 179
Parker, H. S., 51, 55
Peri, J. B., 189
Peterlin, A., 95
Petermann, J., 126
Phelp, D. K., 25
Picken, L., 62
Pickering, H. L., 185
Pieranski, P., 88
Poincaré, H., 63
Poltorak, O. M., 196, 197
Popova, N. M., 194
Preedy, J. E. 95, 126
Price, R. E, 158
Primet, M., 192

Ramquist, L., 180
Randhava, S. S., 190
Rao, P. V., 158
Rastl, P., 181
Rault, J., 86, 88
Read, W. T., 61
Redhead, P. A., 158, 164
Rehmat, A., 190
Reid, A. F., 32
Reid, J. U., 194, 196
Reisman, A., 41
Reneker, D. H., 73
Rhodin, T. N., 162
Richards, W. A., 169
Richardson, G. W., 158
Rideal, E., 185, 186
Rider, J. G., 126
Rieger, E., 165
Ritchie, J. M., 126
Robertson, J. D., 76
Robinson, C., 62
Rooney, J., 192
Rosin, S., 62
Ross, N. D. H., 102
Roth, R. S., 18, 44, 50, 51, 55
Rouse, R. A., 136, 152
Rubin, B. A., 76

Sabatier, P., 184
Sáenz, A. W., 89
Sancier, K. H., 193

Author Index

Sanders, J. V., 3, 5
Saupe, A., 87
Schäfer, H., 41, 61
Scheffer, T. J., 87, 88
Schreiner, D. G., 153, 168
Schuit, G. C. A., 188
Schulte, F., 41
Schwarz, K., 181
Scriven, L. E., 76, 78, 89
Seeger, A., 58
Selwood, P. W., 185
Senderens, J. B., 184
Sermon, P. A., 184, 193
Seto, T., 94, 95
Sheppard, N., 187, 188, 189
Sheridan, J., 190
Shinoda, G., 155
Shirley, D. A., 169
Shockley, W., 61
Siegbahn, H., 169, 180
Silversmith, D. J., 181
Simpson, J. A., 162
Sims, M. L., 165
Sinfelt, J. H., 193
Skinner, H. W. B., 154
Sleeswyk, A. W., 116
Smith, A. E., 185
Smith, D. L., 187
Smith, E., 72
Smith, N. C., 193
Snow, E. C., 170
Sokolova, N. P., 190
Sokol'skii, D. V., 194
Solnyserkova, V. K., 194
Song, S., 95
Spyridelis, J., 14
Staib, P., 155, 180
Starizky, R., 143
Stasny, J. T., 76
Stein, R. S., 95
Steinhaus, H., 75
Steinkilberg, M., 169
Stephens, S. J., 185
Stephenson, N. C., 18, 50, 51

Stephenson, S. T., 164
Stocks, G. M., 178
Straehle, J., 22
Suito, E., 61
Sundberg, M., 15
Suzuki, T., 155
Swanson, N., 177
Swift, C. D., 158
Switendick, A. C., 181
Synge, J. L., 64

Taatjes, S. W., 164
Tan, L. P., 152
Tanaka, K., 94, 95
Taylor, G., 186
Taylor, H. S., 196
Taylor, M. E., 96
Thomas, S., 164
Thompson, M. N., 107
Thomson, S. J., 185, 186, 188, 192, 194, 196
Tilley, R. J. D., 15
Tokonami, M., 56
Tomsett, D. I., 116
Tracy, J. C., 165, 166
Träuble, H., 58
Tripathi, J. B. P., 193
Turner, A. P. L., 126
Turtle, R. R., 157

Ulmer, K., 157, 169
Uyeda, N., 61

van Duijnen, P. Th., 152
van Hardeveld, R., 197
Van Landuyt, J., 14, 21
van Montfoort, A., 197
Vannice, M. A., 193
van Reijen, L. L., 188
van Zeggeren, F., 146
Verbraak, C. A., 116
Verhoeven, J., 165
Verwey, E. J. W., 145
Vescelius, L., 152
Vitek, V., 87

Voigt-Martin, I. G., 131
Volterra, V., 62
Von Dreele, R., 22, 39

Waber, J. T., 170
Walker, D., 143
Wandelt, K., 178
Ward, I. M., 126
Ward, J. C., 62
Ward, J. W., 188
Waring, J. L., 51, 55
Watson, L. M., 156
Wayman, C. M., 99
Webb, G., 184, 185, 186, 187, 188, 192, 193, 194, 196
Weingarten, G., 62
Wells, P. B., 184, 187, 190, 193, 194, 197
Wells, R. G., 59
Wernick, J. H., 179
Wertheim, G. K., 179
Whalley, L., 190
Wheeler, A., 185
Wheeler, E. J., 95, 126
Williams, C., 87, 88
Williams, M. L., 154
Williams, R. W., 178
Williamson, E. D., 151
Wilson, E. B., 66
Wilson, G. R., 194
Wishlade, J. L., 188
Wood, J. C., 146
Wrigley, N. G., 76
Wu, W., 126
Wyrobisch, W., 155

Yamamoto, S., 157
Yasumori, I., 191
Yates, D. J. C., 187
Yee, R. Y., 95
Young, R. J., 126

Zeppenfeld, K., 174

QC
176.8
E4
S 87
v.3

AUG 11 1977